THE
STRANGE
& INFINITE
WORLD OF
NUMBERS

THE
STRANGE
& INFINITE
WORLD OF
NUMBERS

TIM SOLE

ARCTURUS

Acknowledgements

Illustrations by Peter Campbell (http://mr-illustrator.com).

References and recommendations for further reading are:
> The On-Line Encyclopedia of Integer Sequences®
> *The Penguin Dictionary of Curious and Interesting Numbers* by David Wells
> *From Zero to Infinity: What Makes Numbers Interesting* by Constance Reid
> *The Joy of Numbers: The Lexicon of Interesting Integers* by Laurent Hodges
> *Mathematical Puzzles and Diversions* by Martin Gardner
> *More Mathematical Puzzles and Diversions* by Martin Gardner
> www.timsolepuzzles.com

Author Tim Sole is English by birth, has lived in Australia, and is a retired actuary who lives in New Zealand. His hobbies are long-distance cycling (including Athens to Amsterdam, Shanghai to Singapore) and recreational mathematics. Tim has been publishing puzzles since 1980 and three of his titles have won the accolade of being Official American Mensa Puzzle Books. His books have sold more than 600,000 copies.

ARCTURUS

This edition published in 2019 by Arcturus Publishing Limited
26/27 Bickels Yard, 151–153 Bermondsey Street,
London SE1 3HA

ISBN: 978-1-78888-413-6
AD006838UK

Printed in China

Contents ━━━━━━━━━━━━━━━━

Introduction _____

When my weight-lifting daughter completed her first 70 kg snatch I told her it was a 'weird number' (see Glossary). When my mother-in-law turned 85, my birthday message to her was: '85? – That's an A+.' When years ago I bought my first electronic calculator, the first thing I did was to check that 3705, when read upside down on the calculator's screen, said what I thought it would say. Numbers and puzzles have been a longstanding hobby of mine.

Understanding and explaining numbers has also been a large part of my work. In hindsight, I now realize that my passion for puzzles helped my career significantly by widening my problem-solving skills. However, this is not a self-help book! The numbers that make up the 28 chapters in this book are there because they are fascinating – no more and no less.

Fact: 28 is a 'perfect number' (see Glossary), as are 496 and 8128. As well as being perfect numbers, 28 is the sum of the cubes of 1 and 3, 496 is the sum of the cubes of 1, 3, 5 and 7, and 8128 is the sum of the cubes of 1, 3, 5, 7, 9, 11, 13 and 15.

For example, taking the first two chapters of this book, who would have guessed that just over 30 per cent of share prices begin with the digit one (whatever currency they are measured in) or that music relies on a very special mathematical property of a single number – 1.059

(to three decimal places)? And from the last chapter: Georg Cantor's amazing discovery that there are an infinite number of numbers bigger than infinity.

Writing has also been a hobby. My first book was published in 1988 and this will be my tenth. It was intended to be a retirement project that I thought would take many years of happy research. Not so. Such is the richness of numbers, this book felt like it almost wrote itself. The first draft took three months while I was still in full-time work and its many re-writes another six months.

'Re-writes'? Unfortunately yes. In my academic record are six fails in GCSE English exams. Thus when people say they don't get numbers I have sympathy for what they are saying. But most people can enjoy numbers if they forget that's what they are doing. For instance, people who think of themselves as non-numeric, but can calculate in their heads the scores needed to finish a 301 or 501 darts game.

> **Fact: starting from the top and going clockwise, the first four numbers on a dart board are 20, 1, 18 and 4. The 20th, 1st, 18th and 4th letters in the alphabet are T, A, R and D – an anagram of DART.**

There is more to numbers than arithmetic, and you don't need to be a mathematician to enjoy recreational maths. Lots of people will enjoy doing things with numbers as long as they don't think of what they are doing as maths. Just ask the many fans of sudoku. Problems with numbers may seem like hard work for some, but playing with numbers, recreational maths, that's just playing.

Included in the book are 34 puzzles, some of which in slightly different forms have appeared in my earlier books. My favourite is puzzle 21, the 'Ticket to Heaven', which was the title of my first book. This title was apparently confusing to a large London book shop when the book was launched; although it was a puzzle book it was shelved in the Travel section! A few of the 34 puzzles are unashamedly challenging, so feel free to go straight to the answers if you want to.

Also included are descriptions of some of mathematics' most elegant results. When reading these sections avoid over-complicating things if you can. A friend of mine who won a Sylvester Medal – an award for achievement in mathematics from the Royal Society; it is a big deal –

remembers being in tears at school because he couldn't understand how the other children found the ten-times table so easy. No tears over these pages please. If you don't understand a sentence, a paragraph or a page, then skip it and blame me and my six fails in GCSE English.

For reader convenience, the chapter titles are in ascending order, but the book does not have to be read in chapter order. Nor does it need to be read in one go – it can be 'dipped in and out of'.

Sir Isaac Newton is often quoted as saying that his work was built on the shoulders of giants, the geniuses who had come before him. That also applies here. The creative thoughts of many, many people over thousands of years were required before this book could come into being. My hope is that by encouraging others through this book to enjoy the strange and infinite world of numbers, I too will have contributed to this incredible tower of knowledge.

Tim Sole

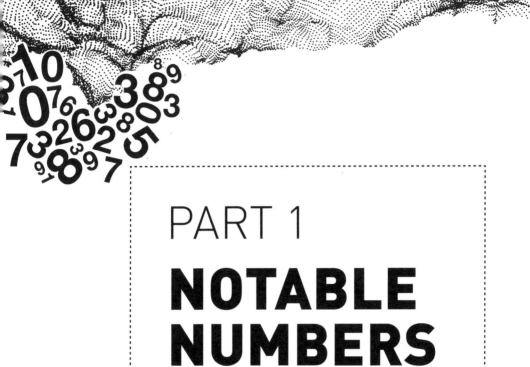

PART 1
NOTABLE NUMBERS

0.301

For every number in this book beginning with a 9 there are seven numbers beginning with a 1. Unusual? An analysis of any newspaper is likely to provide a similar result.

A formula for the percentage of naturally occurring numbers beginning with a specific digit was first suggested in 1881. It appeared in the *American Journal of Mathematics* in a two-page article by Simon Newcomb entitled 'Note on the Frequency of Use of the Different Digits in Natural Numbers'. The article may not sound like a barrel of laughs, but its thesis was intriguing. It predicted amongst other things that the proportion of naturally occurring numbers beginning with a 1 was 30.1%.

Given there are nine digits (ignoring zero), instinctively one would expect the proportion of naturally occurring numbers beginning with a 1 or any of the other digits to be one ninth, which is 11.1%. However, the percentages predicted by Newcomb were as follows:

Digit	Percentage First Digit
1	30.1
2	17.6
3	12.5
4	9.7
5	7.9
6	6.7
7	5.8
8	5.1
9	4.6

Thus according to Newcomb there are as many naturally occurring numbers beginning with a 1 as there are that begin with a 3, 4 or 5 combined, and there are as many naturally occurring numbers beginning with a 2 as there are that begin with a 6, 7 or 8 combined.

Newcomb made his predictions by applying a simple formula using something called logarithms (logs). How he discovered his formula is almost as fascinating as his formula's predictions, so here is a brief history and explanation as to what logarithms are.

In 1614, a Scotsman called John Napier published a book called *A Description of the Wonderful Table of Logarithms*. Napier coined the word 'logarithm' from the Greek words '*logos*', meaning proportion, and '*arithmos*', meaning number. It took Napier twenty years of calculations to produce these tables, but they have saved others literally millions of years' worth of calculations.

What 'taking the log' of a number does is to provide a new number with a very useful property: if it is added to the log of another number, then the anti-log of the sum will be the product of those two numbers. For example, 6, which is 2 x 3, is the anti-log of log (2) + log (3). Adding numbers is easier for people than multiplying them, so in the days before calculators, logs provided a much easier way for scientists and mathematicians to multiply a series of multi-digit numbers. A similar process can be used to change calculations involving division into calculations that only need subtraction. For example, 6 ÷ 2 can be calculated by taking the anti-log of log (6) – log (2).

Calculating logs and anti-logs is not simple (formulae for logs and anti-logs are in the Glossary), so in the days before calculators people would use log tables to 'look them up'. And that's what inspired Simon Newcomb to write his article. He had noticed that the front pages in log table books, which are for numbers beginning with a 1, tended to be far more worn (specifically he said 'grubby') than the back pages of these books, which are for numbers beginning with a 9.

Not this type of log table

Newcomb's formula for the proportion of naturally occurring numbers beginning with the digit 'N' is log (1 + 1/N). Thus for the proportion of numbers beginning with a 1, the predicted proportion from Newcomb's formula is log (1 + 1/1) = log 2 = 0.301.

The simplicity of Newcomb's formula and the coincidence of Newcomb's formula being based on logarithms and being inspired by books of logarithms together seemed too improbable to be true. This, a lack of a formal proof and the formula's counter-intuitiveness was probably why little was done with Newcomb's paper at the time.

In 1937, Frank Benford independently rediscovered the same phenomena. In a paper called 'The Law of Anomalous Numbers' Benford wrote, 'the pages containing the logarithms of the low numbers 1 and 2 are apt to be more stained and frayed by use than those of the higher numbers 8 and 9'. Benford wrote this unaware of the work that Newcomb had done 56 years earlier.

Benford got the 'naming rights' to Benford's Law because he took the trouble to test his theory with actual data – the cold water of a reality check. Using 20,000 naturally occurring numbers from sources such as *Reader's Digest* articles, tables of specific heats of chemical compounds, and population sizes he was able to show the data supported the formula of log (1 + 1/N).

Benford had evidence but no proof. In 1961 a mathematician called Roger Pinkham took a step closer to finding a proof by showing that:

> If there was to be a 'Law of Anomalous Numbers', then that law would have to work regardless of whether the decimal system of counting (the name given to counting in units, tens, hundreds, thousands, etc.) was being used and regardless of which units of measure were employed,

and

> The only formula that could do this was the one that had been suggested by Simon Newcomb and Frank Benford.

In setting this out, Pinkham's paper, 'On the Distribution of First Significant Digits', expressed Benford's Law in a slightly different way. He said that if you take a large group of naturally occurring numbers, then the expected proportion of those numbers that will begin with a digit of N or less is log (N + 1). This gives these results:

Start Digit	Percentage	Start Digit	Percentage
1	30.1	9	4.6
1 or 2	47.7	8 or 9	9.7
1 to 3	60.2	7 to 9	15.5
1 to 4	69.9	6 to 9	22.2
1 to 5	77.8	5 to 9	30.1
1 to 6	84.5	4 to 9	39.8
1 to 7	90.3	3 to 9	52.3
1 to 8	95.4	2 to 9	69.9
1 to 9	100	1 to 9	100

No wonder those log tables had the grubby pages at the front! The 47.7% of naturally occurring numbers beginning with a 1 or a 2 is also the percentage of naturally occurring numbers beginning with the digits 3 to 8 combined. The percentage of naturally occurring numbers beginning with a 9 is just 4.6%.

Benford's Law was finally proved in 1996 by Theodore Hill, an American maths professor.

Many have since tested Benford's Law empirically and found that it also applies, for example, to data sets comprising share prices and land areas and, as Roger Pinkham predicted, that it doesn't matter what currency the share prices are in or which units are used for measuring the land areas.

The 'N' in Newcomb's and Benford's formula does not have to be a single digit, so if you want the proportion of naturally occurring numbers that start with '15' or '123' for example the formula still works. Thus with a little bit of work we can generate the table below for the frequency of digits 0 to 9 occurring as the second digit:

Digit	Percentage Second Digit
0	12.0
1	11.4
2	10.9
3	10.4
4	10.0
5	9.7
6	9.3
7	9.0
8	8.8
9	8.5

Benford's Law has practical applications. For example, it is used for testing for accounting frauds. If someone has created a lot of false invoices without being aware of the Benford patterns and thought they were being clever by spreading the amounts of the false invoices roughly equally over the range of first digits, then a 'Benford test' may highlight the need to carry out a more detailed investigation. Similarly with election results that come from a large number of individual polling booths. Other uses of Benford's Law are measuring the goodness of fit of mathematical models and for allocating data storage on computer disks more efficiently.

1.059 _____

To three decimal places, 1.059 is the fundamental constant of music. If this number did not have the magic qualities it has, then none of the songs that you love would exist. Truly.

The standard piano keyboard as shown in the diagram on the right has 88 notes, comprising seven octaves (more about what this word means later) plus an extra three notes. The note on the far left of the keyboard is called 'A'. The white keys are then called 'B', 'C', 'D', 'E', 'F' and 'G' before restarting at 'A'. Thus there are eight notes on a standard piano keyboard called 'A' and another eight called 'C'. The 'C' nearest the middle is called 'middle C' or 'C4' and the 'A' above that is more precisely known as 'A4'. (That numbering relative to C4 is a little odd, but it is what it is.)

There is no musical significance to a piano having some notes that are black and some that are white. It is just done that way to make the piano easier to play. The notes from left to right are in ascending order of pitch, so the 88th note on the far right of the keyboard, which is C8, is the highest note on this instrument.

Sound is caused by vibration. The frequency of that vibration is measured in cycles per second, or 'hertz', which is commonly abbreviated to 'Hz'. Once one note is set, the others can be tuned to it, but it was not until 1955 that an international standard for pitch was set. The standard is A4 equals 440 Hz, but even today there are still quite a few leading orchestras who set A4 at something other than 440 Hz. For example, the Boston Symphony Orchestra sets A4 at 441 Hz, the New York Philharmonic at 442 Hz, and many European orchestras are setting A4 at 443 Hz.

When a string under tension vibrates it does so in a motion that flows from end to end. The motion is that of a wave, with individual portions of the string making the shape of the letter 'S'. The biggest 'S' is the length of the string, but overlaying that S will be two others of half size running to and from the middle of the string. Those two halves will also have halves and there may be other S-shaped waves that are, for example, a third or a fifth of the string's length. The mix of these extra, secondary notes varies by instrument, which is why a given note played on a violin, for example, sounds the 'same but different' to the equivalent note played on a saxophone.

Because they are in harmony with each other, we call these extra notes harmonics and because of the physics of sound, these harmonics as measured in hertz will be in a simple ratio to the underlying note. Thus whether 440 Hz (A4) is being played on an organ, guitar or oboe, there is going to be some 880 Hz as well. And what is 880 Hz? It is the note one octave higher, A5.

An octave is the musical distance between an 'A' on the keyboard and the 'A's either side of it and it is the same for any other note. An octave, including the notes at each end, contains thirteen notes. The musical gap between notes is called a semi-tone, so notes that are an octave apart are twelve semi-tones apart. Thus the semi-tone ratio is the number that multiplied by itself twelve times equals two, which to six decimal places is 1.059463.

The following table shows the progress in hertz of the 13 notes in an octave:

Number of Semi-Tones From Start	Hertz Multiplier From Start	Multiplier as a Ratio (Approximate)
0	1	1:1
1	1.059463	17:16
2	1.122462	9:8
3	1.189207	6:5
4	1.259921	5:4
5	1.334840	4:3
6	1.414214	7:5
7	1.498307	3:2
8	1.587401	8:5
9	1.681793	5:3
10	1.781797	9:5
11	1.887749	15:8
12	2	2:1

In the table above, if the first note was 'C', then the unshaded notes would be 'C', 'D', 'E', 'F', 'G', 'A', 'B' and 'C', i.e., the scale of 'C major'. The shading is the pattern used by a piano keyboard, with white representing the white notes and black the black notes.

There are four semi-tones from 'C' to 'E' (a musical 'third'), five semi-tones from 'C' to 'F' (a musical 'fourth'), seven semi-tones from 'C' to 'G' (a musical 'fifth'), nine semi-tones from 'C' to 'A' (a musical 'sixth'), and twelve semi-tones from 'C' to 'C' (an octave). These are attractive musical combinations because within the scale of 'C major' and relative to 'C', their ratios are respectively very close to 5:4, 4:3, 3:2, 5:3 and (exactly) 2:1, and the harmonics of these notes overlap in pleasing and sympathetic ways. Indeed, we have a name for this: we call it 'musical'.

It is stunning that within the error margin of the human ear the musical ratios using multiples of 1.059463 work so brilliantly and that they do so regardless of the note on which the scale begins (in music, 'the key'). To get a better fit would need 54 notes in an octave, and humans don't have enough fingers to play the instruments that would then be needed.

A4 set at 440 Hz is known as 'concert pitch', and under concert pitch C4 is 261.6 Hz. 'Science pitch' or 'Saveur pitch', which was proposed in 1713 by Joseph Saveur, set C4 at exactly 256 Hz, thus making C3 = 128 Hz, C2 = 64 Hz and C1 = 32 Hz, which are nice round numbers. It did not catch on.

There is more to music than a sequence of notes and their harmonics. There is also the rhythm and fluctuations in volume that follow their own mathematical patterns. That mathematicians tend to be good musicians and vice-versa is not a coincidence. Gottfried Leibniz, a German mathematician and philosopher said: *'Music is the pleasure the human mind experiences from counting without being aware that it is counting.'*

Puzzle 1: Each square below represents a musician, rock or pop group that has had significant success in the music charts. How many can you work out?

ME$_{\text{ICA}}$	**BGGG**
Ms Wasabi **Ms Paprika** **Ms Turmeric** **Ms Cinnamon**	**Sheep (f)** **Sheep (f)**
$\dfrac{\text{Par}}{2} \quad \dfrac{\text{Par}}{2}$	**EEMM**

The answers to Puzzle 1 are on Page 24.

1.4142135 ————————————

For their second public offering of shares in 2005, Google made available 14,159,265 shares. This number is the first eight decimal places of pi (see page 37). For Google's first public offering of shares on 19 August 2004, the number of shares made available was 14,142,135.

In the table below, Column 2 is Column 1 multiplied by itself. The rule for picking the extra digit for each new row in Column 1 is to make the corresponding entry in Column 2 as big as it can be without exceeding two.

Column 1	Column 2
1	1
1.4	1.96
1.41	1.9881
1.414	1.999396
1.4142	1.99996164
1.41421	1.9999899241
1.414213	1.999998409369
1.4142135	1.99999982358225

Will this table eventually have a row where the number in Column 2 is exactly two or, to ask the question in another way, is there a definitive decimal number, perhaps with a repeating decimal (see next paragraph), that when multiplied by itself will equal exactly two?

To represent a repeating decimal, a notation mathematicians use is to put a dot over the first and last repeating digits. Thus a third for example, which is .3333... (the three dots at the end meaning 'and so on'), is written as $0.\dot{3}$. Other examples are: one sixth can be written as $0.1\dot{6}$, one seventh as $0.\dot{1}4285\dot{7}$ (for more on the number 142857 see page 103), one ninth as $0.\dot{1}$, and one eleventh as $0.\dot{0}\dot{9}$.

Three thirds is one of course, but 0.3 multiplied by three is 'only' 0.9. Can we be sure that a decimal point followed by an infinite number of nines is **exactly** equal to one? The proof is as below:

> 0.9999... x 10 = 9.9999...
> Subtract 0.9999... from both sides
> 0.9999... x 9 = 9
> Thus 0.9999... = 1

In the language of mathematics, the number 1.4142135... is the 'square root' of 2 because 1.4142135... x 1.4142135... = 2. The symbol for square root is written as $\sqrt{}$, so for example $\sqrt{49}$ is 7 because 49 = 7 x 7. Similarly, we say that the 'square' of 1.4142135..., which mathematicians write as $(1.4142135...)^2$, is 2. Sometimes instead of the word 'square' mathematicians say, 'to the power of two' – see the box below.

> Powers provide mathematicians and scientists with a very useful notation: thus 'five to the power of two' is 5^2 = 5 x 5 = 25, 'seven to the power of three' is 7^3 = 7 x 7 x 7 = 343, and 'ten to the power of four' is 10^4 = 10 x 10 x 10 x 10 = 10,000. Where the power is a negative number such as in 10^{-4} for example, then this means 1 divided by 10^4. Thus 10^{-4} = 1 ÷ 10^4 = 1 ÷ 10,000 = 0.0001 and 2^{-2} = 1 ÷ 2^2 = 1 ÷ 4 = 0.25.

The terms 'square' and 'square root' are named with good reason. The area of a rectangle is its height multiplied by its width. Thus the area of a square with sides each measuring three units, for example, is three times three: hence the use of the term 'three squared' for three times three, 'four squared' for four times four, and so on.

Now comes the conundrum that gave Pythagoras (approx. 570 BC to 495 BC) and his followers their big headache. If we take two squares of unit length (i.e., of side 1) and cut them in half along a diagonal we can rearrange them to make a new square as shown opposite:

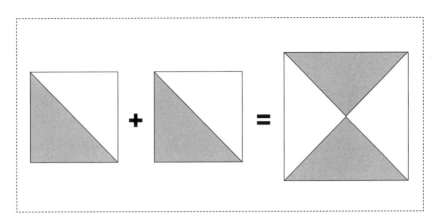

The new square has an area equal to the sum of the two smaller squares, so has an area of 2. Thus the length of the sides of this new square and therefore the length of the diagonals in the smaller squares is a number that when multiplied by itself is 2, i.e., √2, which is 1.4142135…

The ancient Greek philosophers did not know about decimals, but they were aware that two squares of unit measure could be rearranged into one square and it troubled them. This was because they could describe √2 and prove that it was the length of the diagonal in a unit square, but they could not express it as a number.

Pythagoras and his followers expected all numbers that were not integers (whole numbers) to be able to be expressed as one integer divided by another. This seems a reasonable assumption; if with an infinite number of integers to choose from you couldn't express √2 as one integer divided by another, then apparently an infinite set of numbers is not enough! The language of modern mathematics seemingly concurs: numbers that can be expressed as the ratio of two integers are called 'rational' and those that cannot such as √2 are called 'irrational'.

We don't know the proof employed by Pythagoras to show that √2 is irrational, but it is thought to have been similar to that used overleaf. The proof seems complicated first time through, but if the Greeks could do it 2,500 years ago, you can too. The effort is worth it as the proof is ingenious.

Let the integers that can be used to express √2 as a ratio be A and B.

As A / B = √2, then A = √2 x B and A x A = 2 x B x B.

As A and B are whole numbers, then A must be divisible by 2. This means that A can be expressed as 2C where C is also a whole number.

So A x A = 2C x 2C = 2 x B x B, from which 2 x C x C = B x B.

Following the same steps as above, B can be expressed as 2D where D is also a whole number and then we have C x C = 2 x D x D.

These steps can be repeated in perpetuity, but that would not make sense: an integer cannot be divided by two an infinite number of times and still be an integer. Thus we (and the ancient Greeks) conclude that √2 cannot be expressed as the ratio of two integers and therefore that √2 is irrational.

Both questions in Puzzle 2 below use a square root symbol in their answer. They are not easy!

Puzzle 2:

a) With just two twos and no other numbers, find an expression that is equal to three.

b) With just two twos and no other numbers, find an expression that is equal to five.

For your solutions you may only use the mathematical signs that appear in this chapter. The answer to part a) of Puzzle 2 is on page 27 and the answer to part b) is on page 32.

Answers to Puzzle 1:

Metallica, Bee Gees, Spice Girls, U2 (Ewe x 2), Eagles (2 under par in golf is an eagle) and Eminem.

1.618034

This number, known as the 'golden ratio' or the 'divine proportion', is often written as Φ (the Greek letter phi). It is one that is enjoyed by both mathematicians and artists. Two simple equations give a hint as to why mathematicians for over 2,500 years have been so fascinated by it:

2.618034 ÷ 1.618034 = 1.618034 = 1 ÷ 0.618034

Architects, artists and photographers are said to like Φ because a rectangle drawn in the proportion of the golden ratio is said to be the most pleasing to the eye. Similarly with a line divided in the proportion of the golden ratio such as in a cross as shown below, where the lower section of the vertical bar is 1.618 times the length of the upper section.

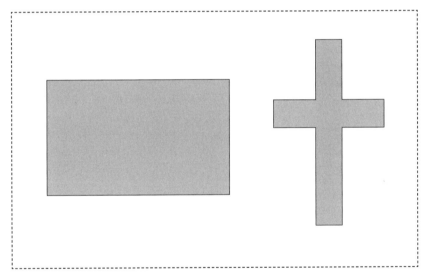

A rectangle and a cross in the proportions of the golden ratio.

Five-pointed stars and pentagrams also feature the golden ratio in their construction, a fact not lost on the Pythagorean School of Greece whose symbol was a five-pointed star in a pentagram. Some 200 or more years later another Greek, Euclid, born around 330 BC, included the golden ratio in his 'Elements', a collection of 13 books on geometry. The golden ratio was an important number for the Greeks, who are widely considered to have deliberately incorporated it into their architecture, the Parthenon being just one example.

Artists whose paintings feature the golden ratio include Salvador Dali (The Sacrament of the Last Supper), Michelangelo (Sistine Chapel), Leonardo Da Vinci (many) and Piet Mondrian (many).

Φ has been calculated to ten trillion decimal places using software called y-cruncher created by Alexander J. Lee. Why? In his own words … 'from a high-school project that went a little too far'. More seriously, it is a way of benchmarking computers.

No discussion on Φ is complete without reference to Fibonacci. It begins with a puzzle posed by Fibonacci in his book *Liber Abaci* (the book of calculation) published in 1202. The book's primary aim was to promote Indo-Arabic numbers (the numbers we use today) over Roman numbers (described on page 73).

This is the puzzle:

> **Start with two baby rabbits, one male, one female. At one month old they mate and one month later they produce two baby rabbits, one male and one female. One month later, at the end of month three, the first pair produces another pair of rabbits. At the end of month four the first pair produces another pair of rabbits and now that they are old enough, so do the second pair. If this pattern continues, how many pairs of rabbits will there be after twelve months?**

The sequence develops as follows: 1 (at month zero), 1, 2, 3, 5, 8, 13, 21, 34, 55, 89, 144, 233, so the answer is 233 pairs of rabbits. Not an easy problem from first principles and even harder to solve if using Roman numbers, but there is a short cut! The numbers in the series, known as the Fibonacci numbers, from the third term onwards are the sum of the two numbers before it, thus 233 = 89 + 144.

And the relevance to Φ? To six decimal places 233/144 = 1.618056, and the further down the sequence one goes the closer the ratio of adjacent terms gets to Φ with successive terms of the series alternating between being too high and being too low. To six decimal places the 17th term, 1597, divided by the 16th term, 987, is 1.618034.

Fibonacci Facts

> Drones (male bees) as well as rabbit pairs have a family tree that follows the Fibonacci series. Drones only have one parent as they hatch from an unfertilized egg. Their mother, the queen bee, who hatched from a fertilized egg, has two parents. Thus the numbers in each generation of a drone's family tree are 1, 1, 2, 3, 5, 8, …

> The ratio of one mile to one kilometre is 1.609 and of a kilometre to a mile is 0.621. Thus consecutive terms of the Fibonacci series are a reasonably accurate way of converting miles to kilometres and vice versa.

> Other than 1, 8 and 144, every Fibonacci number has a prime divisor that is not a prime divisor of any earlier Fibonacci number. 1, 8 and 144 are the only three Fibonacci numbers that are squares or cubes. (For the definitions of prime, squares and cubes see the Glossary/Index.)

> Fibonacci numbers follow Benford's Law (see pages 14 to 16). For example, for the first 400 Fibonacci numbers, 121 (30.2%) of them begin with a 1 and just 18 (4.5%) of them begin with a 9. The predictions for these percentages from Benford's Law are respectively 30.1% and 4.6%.

A curiosity: The reciprocal of 98.99 begins with the Fibonacci series:

$$\frac{1}{98.99} = .01\ 01\ 02\ 03\ 05\ 08\ 13\ 21\ 34\ 55 \ldots$$

The Fibonacci numbers have their own magazine. It is called *Fibonacci Quarterly*, and it has been published at least four times a year since February 1963.

Answer to Puzzle 2a: The solution is $\sqrt{(2 \div .\dot{2})} = 3$. By way of explanation:

$.\dot{2}$ is .2222…, which is $2 \div 9$

$(1 \div .\dot{2})$ is therefore $9 \div 2$

$(2 \div .\dot{2})$ is therefore 9 and $\sqrt{(2 \div .\dot{2})} = 3$

1.96

196 was the number of Prussian soldiers who were accidentally killed by horse kicks from 1875 to 1894. That may seem an old, irrelevant and obscure statistic, but it is one that has a special place in the history of statistics. The number 1.96 for a different reason is also an important number to statisticians.

The human race has been measuring and recording things for thousands of years. The word cubit, the distance from the tip of an adult man's outstretched middle finger to his elbow, is around 5,000 years old for example. The English measure of a 'yard' is less than 2,000 years old and its origin uncertain. It could over time have been defined as any, none, or all of two cubits, a man's girth, or the distance from King Henry 1's nose to the tip of his outstretched fingertips.

The metric system was introduced in France by law in 1795. Three methods were considered for defining a metre: the length of a pendulum that takes exactly one second to swing from one side to the other and back again, one ten-millionth of the distance from the equator to the North Pole, and a physical rod to be held in a secure place whose length would be the definition of one metre.

The last of these is not as simple as it may sound: the length of the rod will vary according to temperature, atmospheric pressure and how it is held (for example, if the rod is held horizontally supported only at each end, then there will be a degree of bowing that will affect

the end-to-end distance). Nevertheless, the physical rod was how the standard was set. The irregularity of the Earth's roundness makes the other two methods even less practical than they may first appear.

With an increasing need for a more precise standard, the metre was redefined in 1960 as a multiple of the orange-red wavelength of krypton-86 in a vacuum. The standard was amended again in 1983 to be the distance travelled by light in a vacuum in 1/299,792,458th of a second. The practicality of accurately measuring the latter is mind-boggling, but what can be done is to measure the time taken for light to travel over a much longer distance and then adjust the answer proportionally.

Aside from the difficulties of defining exactly what a metre is, there are also practical problems with how distance is measured. Today, a yard (36 inches) is defined to be exactly 0.9144 metres, so to two decimal places a metre is 39.37 inches. However, if a large group of people were individually asked to measure to two decimal places a metre bar in inches, it is very unlikely that they would all record a result of 39.37 inches. It is more likely that some would, that many would be quite close, and a few would be not so close.

From empirical evidence it is known that with large samples the average of these measurements will be very close to the actual number (the 'wisdom of crowds'). It is also known that the proportion recording a measure that is over the actual number tends to be very similar to the proportion recording less. The typical pattern is that the measurements tend to cluster around the average with the readings that have the largest errors symmetrically tapering away to a very low proportion. In graph form we get what is known as a 'Normal distribution' (because it is so common), which also because of its shape is sometimes called a bell curve. There is an example below:

A hypothetical distribution of the measurement of one metre when measured in inches by many different individuals is pictured below:

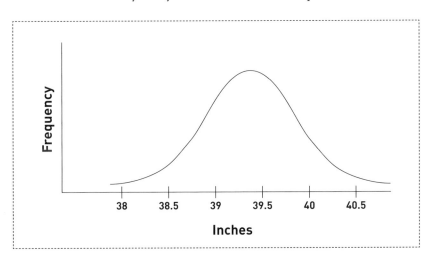

The appearance of graphs such as this depend on the scale used. For example, if the scale was in units of one-hundredth of an inch, then the graph is likely to be stubby and spread out; if the scale is in multiples of whole inches, then it is likely to be tall and skinny. How do we standardize? Using a formula that adjusts for scale we can calculate a 'standard deviation'. Then, on the assumption that the data is 'normally distributed', we would expect on average for 95% of the measurements to lie between 1.96 standard deviations to the left of the mid-point and 1.96 standard deviations to the right. The lower and upper bounds of such a range are sometimes called the 'confidence limits'.

How is this useful? Let us suppose in our metre measuring data that the standard deviation is 0.5 inches. We then expect no more than 2.5% of the results to be less than 39.37 - 0.5 x 1.96 = 38.39 inches and no more than 2.5% of the results to be more than 39.37 + 0.5 x 1.96 = 40.35 inches. Any result outside of this range, 38.39 to 40.35 inches, suggests with 95% confidence that something beyond a random measuring error has happened. It would not prove it of course, random means random, but we would be entitled to be suspicious.

If we change the calculation from 1.96 standard deviations to 2.58 standard deviations, then the range should be capturing 99% of the results and at 3.89 standard deviations either side of the mid-point, the range should be capturing 99.99% of the results. Although we can never

Puzzle 3 is a statistical puzzle. How likely is it that the next president of the USA will have more than the average number of legs?

The answer is on page 36.

be 100% certain that a supposedly random result is not actually random, by using statistical tables we can assign probabilities to results and/ or the average of those results falling outside of our chosen confidence limits. Used properly, these calculations can tell us how likely it is that any given result is just an aberration or whether it is 'statistically significant'.

There are other distribution types apart from the Normal distribution. An example is the Poisson distribution, which is named after the French mathematician Simeon Poisson (1781–1840). This distribution can apply to data where the number of measured events is always a whole number and the measured events are independent of one another. Perhaps the most famous example of a Poisson distribution is the one referred to at the beginning of this chapter and detailed below.

From 1875 to 1894 there were 196 Prussian soldiers from 14 cavalry corps killed accidentally by horse kicks. Ladislaus Bortkiewicz (1868–1931) analysed this data for his book published in 1898 called the *Law of Small Numbers*. For his analysis Bortkiewicz used data from only ten of the 14 Prussian cavalry corps because, he said, the other four corps had a different composition. The table below compares the actual number of horse kick deaths per year over 20 years experienced by ten cavalry corps (an effective sample size of 200) with that predicted by the Poisson distribution:

Number of deaths per cavalry corps per year	0	1	2	3	4	5+
Number in sample with this number of deaths	109	65	22	3	1	0
Number predicted by Poisson distribution	108.7	66.3	20.2	4.1	0.6	0.1

This data fits the Poisson distribution so well that some statisticians questioned whether Bortkiewicz used the Prussian cavalry data selectively, but Bortkiewicz refuted this.

The number of goals scored in football matches is another often quoted example of the Poisson distribution. If you know how many goals a football team scores on average, then assuming the Poisson distribution you can work out the expected percentage of games where they score no goals, one goal, two goals, three goals, etc. The example below is for a team that scores on average 1.96 goals per match.

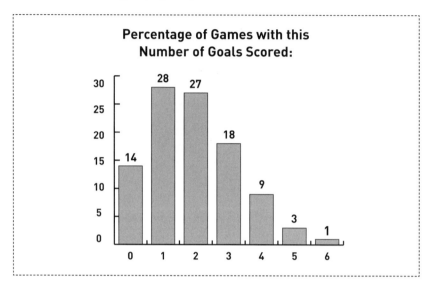

Percentage of Games with this Number of Goals Scored:

A statistician's role is to find the theoretical distribution that fits the data the best – there are tests for goodness of fit – and amend the theoretical distribution or distributions used accordingly in order to make the best tool for making predictions. In the bar chart above for example, it may be that a better fit (and predictor of results) would be obtained by using different averages for goals scored playing away and goals scored playing at home. This is the type of analysis that a statistician who works for a betting agency might carry out.

Answer to Puzzle 2b: The solution is $\sqrt{(.2)^{-2}} = 5$.

By way of explanation:

$(1 \div .2) = 5$

$(1 \div .2)$ can be written as $(.2)^{-1}$

$(.2)^{-1}$ can be written as $\sqrt{(.2)^{-2}}$ and therefore $\sqrt{(.2)^{-2}} = 5$

2.71828

2.71828... is sometimes known as Euler's number. Leonhard Euler (1707–1783) is one of the greatest and most prolific mathematicians in history. In the words of the mathematician Pierre-Simon Laplace (1749–1827), who is sometimes referred to as the Isaac Newton of France: 'Read Euler, read Euler, he is the master of us all.' The symbol worldwide for Euler's number is e.

There are many formulae that involve Euler's number; an example is that e equals the sum of the reciprocals of all of the factorials from zero to infinity (see page 54). Two more are the formulae for the Normal and Poisson distributions described in the previous chapter. Below is an illustration of another:

2/1	= 2.0
3/2 x 3/2	= 2.25
4/3 x 4/3 x 4/3	= 2.37037
5/4 x 5/4 x 5/4 x 5/4	= 2.44141
6/5 multiplied by itself 5 times	= 2.48832
51/50 multiplied by itself 50 times	= 2.69159
501/500 multiplied by itself 500 times	= 2.71557
5,001/5,000 multiplied by itself 5,000 times	= 2.71801
50,001/50,000 multiplied by itself 50,000 times	= 2.71825

Given the title of this chapter it is not hard to guess where this is going.

If we require a mnemonic to remember e, then it would be hard to better – 'we require a mnemonic to remember e'. It works by counting the letters in each word, thus 'we' = 2, 'require' = 7, etc. The sixth to ninth decimal places of e are 1828, the same as the second to the fifth, but that is just a curious coincidence.

Euler's number is also the basis of something called natural logarithms and its inverse, the natural exponential function (defined in the Glossary), which are fundamental to calculus. The rest of this chapter, however, will focus on a formula called Euler's identity, a formula that mathematicians swoon over. For this we need to define the number called 'i'.

Numbers that can be expressed as the ratio of two integers are called 'rational' (derived from the word 'ratio') and those that cannot be expressed as the ratio of two integers are called 'irrational' (see pages 23–24). Together the rational and irrational numbers make up what is called the 'real numbers'. They are the numbers that when pictured on an infinite line with zero at the centre spread from minus infinity on one side to plus infinity on the other side. Every number from zero to plus infinity is then the measure of its distance from point zero, and the negative of that number is the measure from zero going in the opposite direction.

Describing numbers as an infinite line is to treat numbers as one-dimensional, but what if we said that two dimensions were required to express a single number? (The best mathematicians are good at asking questions like this.) What if instead of flipping 180 degrees to go from plus one to minus one you could do it in two flips of 90 degrees each, and where would you be in our two-dimensional number space after flip one?

Where you would be is on one of two numbers that when multiplied by itself is equal to minus 1. There are two answers because you can choose to flip 90 degrees clockwise or 90 degrees anti-clockwise. Mathematicians call these numbers 'i' and 'minus i', and these two numbers are the two square roots of minus 1. A diagram of two-dimensional number space is shown opposite.

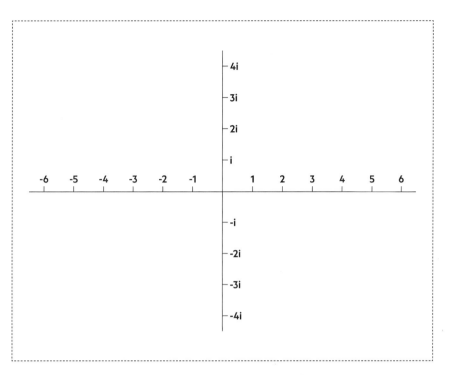

In contrast to the 'real' numbers on the horizontal line in the diagram above, all the numbers on the vertical line are called 'imaginary' numbers. (I am not making this up!) Imaginary numbers were named as such by René Descartes (1596–1650), who is also the author of the famous quote: 'I think, therefore I am.'

Numbers formed by adding real numbers and imaginary numbers are called complex numbers. No need to ask why. Complex numbers are used in mathematics in all sorts of creative ways and notwithstanding the abstractness of the concept, have many useful applications in science.

On page 22 there is a box describing 'powers'. Suffice it to add that powers do not need to be whole numbers, so they can be any type of number including imaginary irrational numbers such as iπ. (See next chapter for a definition of π.)

Below is Euler's identity, where i is the square root of minus 1 and e is Euler's number. This formula is the gold standard for mathematical elegance. In one equation it brings together mathematics' five most famous numbers.

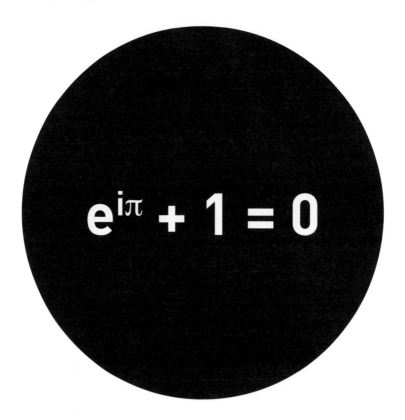

$$e^{i\pi} + 1 = 0$$

Answer to Puzzle 3. Some people have no legs, some only one, and more than two legs is very rare. The average number of legs is therefore slightly under two. Thus it is very likely that the next president of the USA will have more than the average number of legs.

3.14159265

To eight decimal places and irrespective of its size, the ratio of a circle's circumference to its diameter is 3.14159265. Similarly, to eight decimal places and irrespective of its size the ratio of a circle's area to that of a square whose sides are the same length as the circle's radius is also 3.14159265. The two ratios to as many decimal places as you want are identical, and today we know this number as π, pronounced 'pi', the 16th letter in the Greek alphabet. Establishing a numerical value for π to ever greater degrees of accuracy has a long history.

In the British Museum there is a document known as the Rhind Papyrus. It is Egyptian and it is thought to have been written around 1650 BC. It is essentially a maths text book containing 21 problems and their solutions in arithmetic, 20 in algebra, 21 in geometry and 29 miscellaneous. The Rhind Papyrus shows that the Egyptians had an impressive grasp of mathematics including a reasonable estimate for π. Their estimate for π was 256/81 = 3.16049, which is only 0.6% higher than π's actual value.

By around 250 BC the Greeks had an estimate of π that was 99.95% accurate

Year	Number of decimal places
1400	10
1706	100
1855	500
1949	2,037
1958	10,021
1961	100,265
1973	1,001,250
1983	16,777,206
1987	134,214,700
1989	1,073,740,799
1997	51,539,600,000
1999	206,158,430,000
2002	1,241,100,000,000
2011	10,000,000,000,050

(see page 110) and by around AD 480 a Chinese mathematician, Zu Chongzhi, had accurately calculated π to seven decimal places. He also discovered that 355/113 is a very good estimate for π (it is correct for six decimal places).

The record for calculating π accurately then progressed as shown on page 37. Calculations from 1949 onwards were done by computer. The record was broken again in 2016 when π was calculated to 22.4 trillion decimal places.

Problem 48 of the Rhind Papyrus invites the reader to compare the area of a circle with that of its circumscribing square. It is the oldest preserved example of the 'how-to-square-a-circle' question; that is, how to construct a square that has exactly the same area as a given circle. After more than 3,000 years of trying it was finally proved in 1882 that it cannot be done. The proof is a consequence of the proof that π is not only irrational, it is also transcendental, which means it can only be expressed in terms of other transcendental numbers or as an equation that requires an infinite number of terms.

As well as occurring in geometry and trigonometry, the number π occurs in several of mathematics' most famous formulae. These include special cases of Riemann's zeta function (page 66), the Normal distribution and Euler's identity. A further example is four minus four thirds plus four fifths minus four sevenths plus four ninths and so on, which adds up to π.

In the early 1970s, a computer magazine ran a Christmas competition for coming up with an easy way for remembering the first few digits of π. Below is the winning entry. The mnemonic works by counting the letters in each word, beginning with the number of letters in the verse's title (PIE).

This tells us that π to 20 decimal places is 3.14159265358979323846. In a similar vein, but offering only the first 15 digits of π, we have this from Sir James Jeans:

How I want a drink; alcoholic of course, after the heavy lectures involving quantum mechanics.

PIE
I wish I could determine pi
Eureka, cried the great inventor
Christmas pudding, Christmas pie
Is the problem's very centre

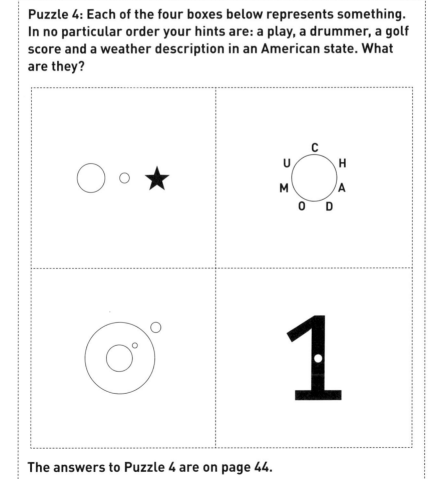

The values of π provided by these mnemonics are far more accurate than is ever needed for day to day living. For example, knowledge of the Earth's diameter and just the first ten digits of π will provide an estimate of the Earth's circumference with an error margin of less than one centimetre.

Knowing π to 20 decimal places is impressive enough, but the certified record by the *Guinness Book of Records* held by Rajveer Meena for reciting π from memory is 70,000 decimal places! The recitation took 9 hours and 27 minutes. Perhaps even more amazingly, Johan Zacharias Dase in 1844 was able to calculate π in his head to 200 decimal places. It took him two months.

Puzzle 4: Each of the four boxes below represents something. In no particular order your hints are: a play, a drummer, a golf score and a weather description in an American state. What are they?

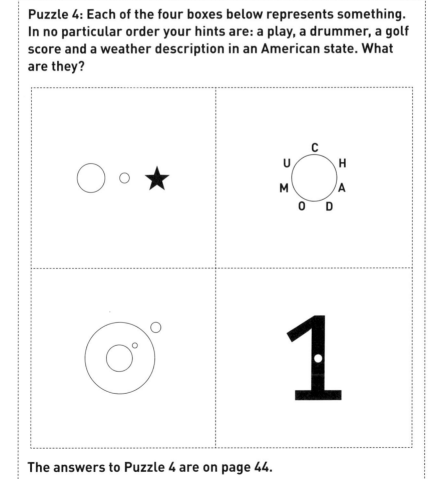

The answers to Puzzle 4 are on page 44.

PART 2
INTERESTING INTEGERS

7

Sitting in a betting shop on Seventh Avenue on 7 July (the seventh of the seventh), a punter notices that horse number seven running in the 7:00 pm race is called Lucky Seven. He searches in his pockets and finds he has $7.77 left. He bets the $7.77 on this horse to win – how could it not he thinks. He thinks wrong: the horse came seventh.

There are seven days in a week, seven ages in Shakespeare's ages of man, Ian Fleming's James Bond was agent 007, there were seven wonders of the ancient world, there are the so-called seven seas, and we say there are seven colours in the rainbow. It is not surprising that seven is often considered to be a lucky number and in Alex Bellos's global online poll to find the world's favourite number, it was number seven that came out on top.

> **Puzzle 5 is an old puzzle. As I was going to St Ives I met a man with seven wives. Seven wives with seven sacks, seven sacks with seven cats, seven cats with seven kittens. How many were going to St Ives? It certainly is an old puzzle; something similar is included in the Rhind Papyrus dated circa 1650 BC.**
>
> **The answer to Puzzle 5 is on page 47.**

Considered by many to be unlucky is the number 13 (and yet in Alex Bellos's poll of favourite numbers it came sixth). In the Western world some hotels have no floor between floors 12 and 14 or, if they do, it might be labelled floor 12a. In some dialects of Chinese the word for four sounds like the word for death, so the number four is considered unlucky and it is not unusual for a Chinese hotel not to have a floor 4. Fear of the number 4 is called tetraphobia and fear of the number 13 is called triskaidekaphobia. Love of the number 13 also has its own name – triskaidekaphilia.

Puzzle 6: The cross-number below was published by 'Reflector' in the UK magazine *The Listener* under the name of *'Lucky Numbers'* more than 50 years ago. The author, whose real name was W. M. Jeffree, must have been very proud of it. Based on multiples of seven and thirteen, even the frame measures seven by thirteen.

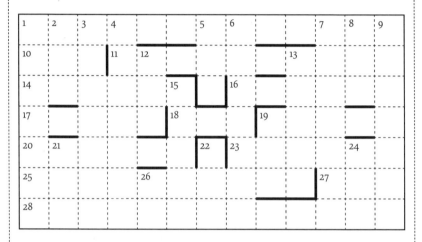

All across clues (1, 10, 11, 14, 16, 17, 18, 19, 20, 23, 25, 27 and 28): divisible by 13

All down clues (1, 2, 3, 4, 5, 6, 7, 8, 9, 12, 13, 15, 19, 21, 22, 24 and 26): divisible by 7

Additional clues:

A. No two answers are the same

B. No answer contains a zero

C. All the answers are palindromes (a number equal to itself when read backwards)

A start for Puzzle 6 in on page 52 and the answer is on page 57.

11

Under the Calendar (New Style) Act 1750, the British Isles, the English colonies and America 'lost' 11 days during September 1752. Literally overnight, the date jumped from 2 September 1752 to 14 September 1752. The change happened because the Julian calendar was replaced by the Gregorian calendar. The Julian calendar, introduced by Julius Caesar, had itself replaced the Roman calendar in 45 BC. The Roman calendar had ten named months, which is the explanation for the appearance of 'septem', 'octo', 'novem' and 'decem', the Latin for seven, eight, nine and ten, in our names for the last four months of the year. It was the Julian calendar that introduced the new months of July (from Julius) and August (from Augustus).

February 2022

Mon	Tue	Wed	Thu	Fri	Sat	Sun
	1	2	3	4	5	6
7	8	9	10	11	12	13
14	15	16	17	18	19	20
21	22	23	24	25	26	27
28						

22/2/22 – It's Twos Day!

The Julian calendar was based on a year lasting 365 days and 6 hours, with the extra 6 hours being achieved through adding a leap day every fourth year. However, the solar year (the time for the Earth to orbit the sun) is 365 days, 5 hours, 48 minutes and 45 seconds, which is 11 minutes and 15 seconds less. To put the religious festivals back to the dates that had been set by the Church in AD 325, the Gregorian calendar (named after Pope Gregory XIII), retrospectively converted the years since AD 325 that ended in '00' that were not a multiple of 400 back to normal years, making 97 instead of 100 leap years every 400 years.

There was a 'loss' of 10 days when the Gregorian calendar was introduced in 1582. Although the extra 11 minutes and 15 seconds per year compared to the Julian calendar may not sound very much, over twelve and a half centuries it clearly builds up. When Russia (1918), Greece (1923) and Turkey (1926) changed to the Gregorian calendar, they each 'lost' 13 days. Since 1972, a very small portion of this time has been given back. Between 1972 and 2016, 27 leap seconds have been added to compensate for the days getting longer because the Earth's rotation on its axis is slowing.

> **Puzzle 7: Cricket teams have 11 players. In one innings of cricket, each of the ten batsmen out was clean bowled first ball. Which numbered batsman was not out?**
>
> **The answer to Puzzle 7 is on page 53.**

> **Puzzle 8: For readers raised west of the Atlantic Ocean and not familiar with the rules of cricket, this puzzle is for you. In which states are the southernmost part of the USA, the northernmost part, the easternmost part and the westernmost part?**
>
> **The answers to Puzzle 8 are on page 57.**

In Britain and Commonwealth countries, 11 November is named Armistice Day to mark the armistice signed between the Allies of World War I and Germany at 11:00 a.m. on 11/11/1918. In China, 11 November is 'singles day'. Created in response to Valentine's Day, the date was chosen because '11' is reminiscent of bare branches, a Chinese expression for bachelors and spinsters.

There is a quick way of determining whether a number is divisible by 11: if the sum of every other digit in a number is the same as the sum of the rest of the digits it will be divisible by 11. For example, 1,243 is divisible by 11 as the first digit plus the third digit (1 + 4) equals the sum of the second and fourth digits (2 + 3). Similarly, 28,765 is divisible by 11 as 2 + 7 + 5 = 8 + 6.

The above is a special case of the general rule that if the sum of the digits in the odd-numbered positions (first, third, fifth, seventh, etc.) of a number less the sum of the digits in the even-numbered positions of that number is divisible by 11, then the number will be divisible by 11. Thus 808,082 is divisible by 11 because 8 + 8 + 8 − 2 = 22, which is divisible by 11.

Puzzle 9: It is almost unheard of for a chess grandmaster to be mated in just eleven moves, but it happened in Vienna in 1910. Reti, playing white and also a grandmaster, checkmated Tartakower on his eleventh move. Below is what the game looked like after Black's eighth move:

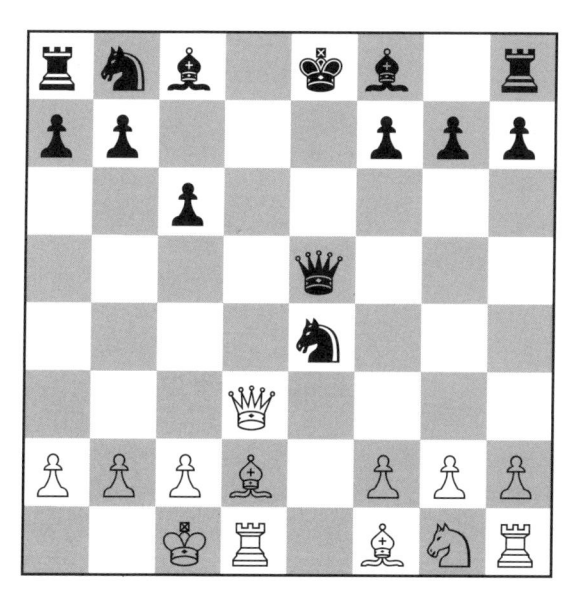

White to play and mate in three.

The answer to Puzzle 9 and the full game is on page 61.

The answer to Puzzle 9 and the full game is on page 61.

Answer to Puzzle 5. The question is far from precise. For example, are the seven wives travelling with the man and do they have seven sacks between them or seven sacks each? It does not matter as the puzzle is a trick. If 'I was going to St Ives' and met others, it is presumed that they were travelling in the other direction, so only one was going to St Ives – the person asking the question.

15 _____

A magic square has rows, columns and long diagonals that add up to the same number. In the more interesting magic squares the cells are generally all different and follow a simple rule, e.g., between them they contain all of the digits from 1 to 9. An example of a three-by-three magic square is shown below:

6	1	8
7	5	3
2	9	4

The rows, columns and long diagonals of this magic square each add up to 15.

To show that a 5 will always appear in the middle of a three-by-three magic square comprising the digits 1 to 9, consider the middle column, the middle row, and the two long diagonals. These four lines comprise one of each number on the outside of the square and four of the number in the middle. Call this 'Group A', noting that the digits in Group A add up to 15 x 4 = 60.

Let the three rows (or three columns if you prefer) be 'Group B'. The digits in Group B add up to 15 x 3 = 45. Group A minus Group B is the number in the middle three times and is also 60 – 45 = 15. Thus the number in the middle is always 15 ÷ 3 = 5.

In our next magic square the constant is 34 and it is made up from the integers 1 to 16. It is famous for featuring in Albrecht Dürer's engraving 'Melancholia 1'.

16	3	2	13
5	10	11	8
9	6	7	12
4	15	14	1

The rows, columns and long diagonals each add up to 34, as do the four corners, the four two-by-two squares in each corner, and the two-by-two square in the middle. For the finishing touch see the middle digits in the bottom row – the engraving was done in 1514.

The Indian mathematician Srinivasa Ramanujan created a magic square with a constant of 139.

22	12	18	87
88	17	9	25
10	24	89	16
19	86	23	11

Once again the rows, columns and long diagonals each add up to the same constant, as do the four corners, the four two-by-two squares in each corner, and the two-by-two square in the middle. What makes it special to Ramanujan is that the top row is Ramanujan's birthdate, 22/12/1887.

Magic hexagons are also possible:

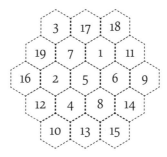

This one's constant is 38.

The magic square overleaf is still magic when each of its elements is squared (multiplied by itself), increasing the constant from 260 to 11,180.

7	53	41	27	2	52	48	30
12	58	38	24	13	63	35	17
51	1	29	47	54	8	28	42
64	14	18	36	57	11	23	37
25	43	55	5	32	46	50	4
22	40	60	10	19	33	61	15
45	31	3	49	44	26	6	56
34	20	16	62	39	21	9	59

The magic square below is made up of the first 144 odd primes allowing 1 as a prime:

1	89	97	223	367	349	503	229	509	661	659	827
823	83	227	653	379	359	523	491	199	101	673	3
821	211	103	499	521	353	233	373	73	643	677	7
809	79	107	197	383	647	337	487	541	239	683	5
811	641	193	109	241	389	547	461	347	691	71	13
797	631	557	113	467	331	397	251	191	701	67	11
19	619	719	563	257	317	421	443	181	127	61	787
29	709	727	479	263	311	17	463	569	131	47	769
313	617	607	173	269	409	401	137	577	179	59	773
31	53	139	761	167	307	271	439	571	613	743	419
23	43	757	587	601	293	431	457	163	277	733	149
37	739	281	157	599	449	433	283	593	151	41	751

The rows, columns and long diagonals each add up to 4,514. The finishing touch to this magic square is that its two diagonals from left to right are in ascending order.

Returning to the magic square that began this chapter, which was:

6	1	8
7	5	3
2	9	4

Consider three dice, A, B and C. The faces on dice A have the numbers on the first row of the magic square repeated, so are 6, 6, 1, 1, 8 and 8. Dice B has the numbers on the second row of the magic square repeated, and dice C has the numbers on the third row of the magic square repeated.

Although the numbers on each dice are all different, for each dice the sum of the digits is 15 x 2 = 30.

In a game between two players each using one of the above dice where the player throwing the higher number wins:

> Dice A on average will beat dice B five times out of nine.

> Dice B on average will beat dice C five times out of nine.

> Dice C on average will beat dice A five times out of nine.

These three dice are examples of 'nontransitive dice'. Had we used the columns of the three by three magic square for the dice instead of its rows, so the faces on dice A would have been 6, 6, 7, 7, 2 and 2 for example, the result would have been the same: dice A beats dice B, dice

B beats dice C, and dice C beats dice A each on average five times out of nine.

Demonstrated below is another feature of the above three by three magic square. Begin by squaring the numbers formed by the rows left to right and then right to left and compare the totals:

$$618^2 + 753^2 + 294^2 = 816^2 + 357^2 + 492^2$$

And then square the numbers formed by the columns top to bottom and then bottom to top:

$$672^2 + 159^2 + 834^2 = 276^2 + 951^2 + 438^2$$

This also works with the diagonals:

$$654^2 + 132^2 + 879^2 = 456^2 + 231^2 + 978^2$$
$$654^2 + 798^2 + 213^2 = 456^2 + 897^2 + 312^2$$
$$852^2 + 396^2 + 417^2 = 258^2 + 693^2 + 714^2$$
$$852^2 + 174^2 + 639^2 = 258^2 + 471^2 + 936^2$$

That's magic!

A start for Puzzle 6. 26d is a two-digit palindromic number divisible by 7, so is 77. Thus the 6th digit of 25a, the 2nd digit of 15d, the 3rd digit of 18a, and the 9th digit of 28a are all sevens. 18a being divisible by 13 is 767.

2d and 8d are three-digit palindromic numbers that begin with the same digit, are divisible by seven and are different from one another. 2d is therefore 525, 595, 616 or 686. It cannot be 525 or 616 otherwise 10a has no solution.

If we assume 2d is 595 and then apply similar logic to 21d and 24d to that used in the paragraph above for 2d and 8d, then that would lead to 11a being 525444525, which it cannot be as this number is not divisible by 13. Therefore 2d is 686.

Answer to Puzzle 7: The batsmen out in the first six balls are 1, 3, 4, 5, 6 and 7. Batsman 8 replaces batsman 7, but with six balls bowled there is a change of bowler and a change of ends. The wickets taken by the second bowler are those of batsmen 2, 9, 10 and 11. The batsman not out is number 8.

23

23 is the number of people needed for a game of football including the referee. If they walk on to the playing field one after another, in how many different ways can they do this?

There are 23 choices for the first person to walk on to the playing field, leaving 22 still to come. For the second person to walk on to the playing field there are 22 choices, and similarly 21 for the third, 20 for the fourth, and so on. So the answer to the question is 23 x 22 x 21 x 20 x ... x 3 x 2 x 1. In mathematics this is written as 23! and pronounced as 'twenty-three factorial'.

Factorials are useful in statistics because when we have a number of equally likely outcomes, the probability of the event we are interested in will be the total number of ways this event can happen divided by the total number of possible outcomes. Calculating the number of these outcomes often involves the use of factorials.

Factorial numbers become large very quickly and consequently their reciprocals (a reciprocal of a number is one divided by that number) get small very quickly. The table below illustrates:

Factorial	Value of Factorial	Reciprocal of Factorial
1!	1	1
2!	2	0.5
3!	6	0.16667
4!	24	0.04167
5!	120	0.00833
6!	720	0.00139
7!	5040	0.00020
8!	40,320	0.00002

By convention, 0! = 1, so the sum of the reciprocals of the first nine factorials beginning with 0! is 2.71828 – a number in this book that very deservedly has its own chapter heading. See page 33.

The last number in the middle column of the table opposite (40,320) is the number of minutes in four weeks. It is also the number of changes when ringing a full peal of eight bells.

Returning to our footballers, we can use a technique called mathematical induction to prove as shown opposite that 23! is the number of different ways 23 people can walk on to a playing field one after another. There are two steps to this.

For the first step we assume that 22! is the right answer for 22 people and we have those permutations listed. Looking at the first permutation on the list, the 23rd person could go to the front, to the back, or in any one of the 21 gaps, making 23 new permutations. This applies to every one of the 22! permutations on the list, so if 22! is right for 22 people, then the answer for 23 people will be 22! x 23, which is 23!

The next step is crafty. We apply the logic of step one to show that 22! is correct for 22 people if 21! is correct for 21 people, that 21! is correct for 21 people if 20! is correct for 20 people, and so on, all the way down to 3! is correct for 3 people if 2! is correct for 2 people. We know that 2! is correct for two people, because we can count that there are only two different ways that two people can walk on to a playing field one after the other. Thus working back up the chain, 23! is confirmed as the right answer.

Puzzle 10: Problem 23 of the Rhind Papyrus (see page 37) is: If seven reciprocals add up to two-thirds and five of them are the reciprocals of 4, 8, 10, 30 and 45, what are the other two?

The answer to Puzzle 10 is on page 61.

For the International Congress of Mathematicians in Paris in the first year of the twentieth century the German mathematician David Hilbert (1862–1943) presented as a challenge to his colleagues a paper with 23 unsolved mathematical problems. These problems set much of the agenda for mathematical research for the twentieth century. As of today, ten of the 23 problems have been solved, seven have been partially solved, two have not been solved and four are now considered too vague to have a solution.

The Milky Way galaxy is estimated to contain 100 to 400 billion stars and the universe 100 to 200 billion galaxies. An estimate of the number of stars in the universe is therefore between 10 and 80 thousand billion billion, a 23-digit number. Coincidentally, 23! is also a 23-digit number.

The number of people needed in a room for it to be more likely than not that two or more of them share a birthday is 23. This statement often surprises people the first time they hear it because they expected the number to be higher. This is because (a) they probably hadn't considered that 23 people in a room provides 253 (= 23 x 22 / 2) possible pairings and (b) that there are many ways for two or more people in a room to share a birthday: it could be three on the same day, four on the same day, two pairs of shared birthdays, etc.

The probability of two or more people sharing a birthday becomes 71% with 30 people in the room (435 possible pairings), 89% with 40 people in the room (780 possible pairings), and 97% with 50 people in the room (1,225 possible pairings). With 70 people (2,415 possible pairings), it is 99.9%. For the probability of three or more people in a room sharing a birthday to be at least 50%, you need at least 88 in the room (109,736 possible groups of three). Turning the question around, the number of people needed in a room so that there is a 50% chance of everyone sharing a birthday with someone else is 3,064.

Puzzle 11 is a birthday puzzle: How can two babies be born in the same place on the same day in the same week in the same month in the same year with the same mother and the same father and not be twins?

The answer to Puzzle 11 is on page 70.

Answer to Puzzle 6:

1	2	3	4			5	6			7	8	9
5	6	8	5	1	7	9	7	1	5	8	6	5
10			11	12				13				
5	8	5	4	1	4	5	5	5	5	4	1	4
14					15		16					
7	6	7	7	6	7	9	5	6	5	5	6	5
17				18			19					
1	7	7	7	1	7	6	7	2	7	8	7	2
20	21					22	23				24	
7	5	7	7	5	7	3	5	5	5	5	5	5
25			26						27			
5	2	5	4	7	7	4	5	2	5	4	9	4
28												
5	5	8	5	7	7	3	7	7	5	8	5	5

28

There are 28 domino tiles in a standard set of dominoes. The tiles usually show dots rather than numbers, but essentially they look like this:

| 0 | 0 | | 0 | 1 | | 0 | 2 | | 0 | 3 | | 0 | 4 | | 0 | 5 | | 0 | 6 |

| 1 | 1 | | 1 | 2 | | 1 | 3 | | 1 | 4 | | 1 | 5 | | 1 | 6 | | 2 | 2 |

| 2 | 3 | | 2 | 4 | | 2 | 5 | | 2 | 6 | | 3 | 3 | | 3 | 4 | | 3 | 5 |

| 3 | 6 | | 4 | 4 | | 4 | 5 | | 4 | 6 | | 5 | 5 | | 5 | 6 | | 6 | 6 |

With every number from 0 to 6 paired once with itself and once with every other number.

Below shows two boxes made from two sets of dominoes that have been muddled up, but for one set the tiles have all been placed horizontally and for the other set they have all been placed vertically. Each box contains tiles from both sets (so contains a mixture of horizontal and vertical dominoes). As a start, the double zero vertical tile has been marked in.

Puzzle 12: Identify the position of each of the 56 dominoes.

4	2	1	4	0	2	2	3
1	5	6	4	0	6	1	3
3	3	6	2	6	3	5	5
6	0	1	2	4	2	3	4
4	0	3	1	1	1	0	0
3	5	4	3	2	5	3	1
3	6	4	0	2	5	4	5

1	0	5	6	0	3	6	1
1	3	3	4	2	2	4	1
4	6	1	1	6	0	2	6
5	2	5	1	6	2	5	5
4	5	0	0	6	6	0	0
2	5	1	3	5	6	0	3
4	4	0	2	4	2	5	6

The answer to Puzzle 12 is on page 65.

By placing the dominoes such that all the numbers in the top row are zero (usually depicted on a domino as a blank), we can pretend that the top row isn't there and create a seven-by-seven magic square. Here is an example:

0	0	0	0	0	0	0
6	4	1	4	1	5	3
5	2	6	4	1	5	1
5	2	5	2	6	3	1
3	1	4	2	6	2	6
3	6	3	1	4	2	5
2	5	3	6	3	1	4
0	4	2	5	3	6	4

Ignoring the row of zeros at the top, the rows, columns and long diagonals each add up to 24.

Games of dominoes usually involve players taking turns to build up a chain where every new domino placed must match the end of the domino it is laid next to. Below shows a finished game for which the sums of the numbers on the two columns both equal 22 and the sums of the numbers on the two rows both equal 66.

Puzzle 13: Find another rectangle from a finished game using all 28 dominoes where the numbers on each of the four sides of the rectangle add up to 44.

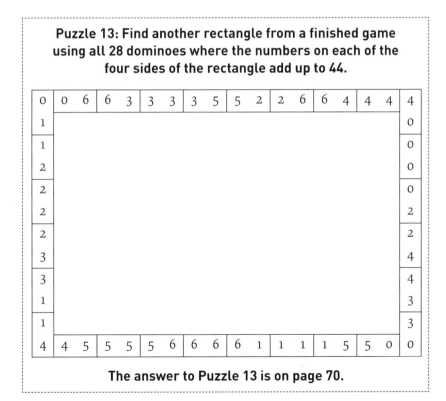

The answer to Puzzle 13 is on page 70.

> **Puzzle 14:** It is easy to place 32 domino tiles to cover the 64 squares of a chess board. If the top left square and the bottom right square of the chess board were removed, could the remaining 62 squares be covered by 31 dominoes?
>
> **The answer to Puzzle 14 is on page 74.**

Other than 1, the divisors of 28 are 2, 4, 7, 14 and 28. Because 1/2 + 1/4 + 1/7 + 1/14 + 1/28 sums to exactly 1, we say that 28 is a 'perfect number'. It was Euclid in around 300 BC who coined (in Greek) the name 'perfect number'.

Another way of defining a perfect number is any number whose divisors including one but excluding itself add up to that number. Thus 28 is a perfect number because 1 + 2 + 4 + 7 + 14 = 28. Perfect numbers are hard to find. Only four were known by AD 100. They were: 6, 28, 486 and 8,128. The eighth, 2,305,843,008,139,952,128, was discovered in the eighteenth century by Leonhard Euler. Through the use of supercomputers the list of known perfect numbers currently has 49 entries.

It is not known whether the number of perfect numbers is infinite and whether there are any odd perfect numbers (the 49 perfect numbers known are all even). So, when you have finished reading this book and you are looking for something to do ...

> **Puzzle 15:** What in English is the smallest integer that cannot be defined in less than 28 syllables?
>
> **The answer to Puzzle 15 is on page 83.**

Answer to Puzzle 10: The other two are the reciprocals of 9 and 40. The reciprocals of 4, 8, 9, 10, 30, 40 and 45 add up to 240/360, which is two-thirds.

Answer to Puzzle 9: One assumes that Black played 8. . . . . N x N?? expecting White to reply 9. R – K1, which would have pinned Black's knight and allowed White to recapture his lost piece. White's reply, however, was much more exciting. The full game was as follows:

RETI	TARTAKOWER
1. P – K4	P – QB3
2. P – Q4	P – Q4
3. N – QB3	P x P
4. N x P	N – B3
5. Q – Q3	P – K4
6. P x P	Q – R4 ch
7. B – Q2	Q x KP
8. 0–0–0	N x N??
9. Q – Q8 ch!	K x Q
10. B – N5 dbl ch	K – B2
11. B – Q8 mate	

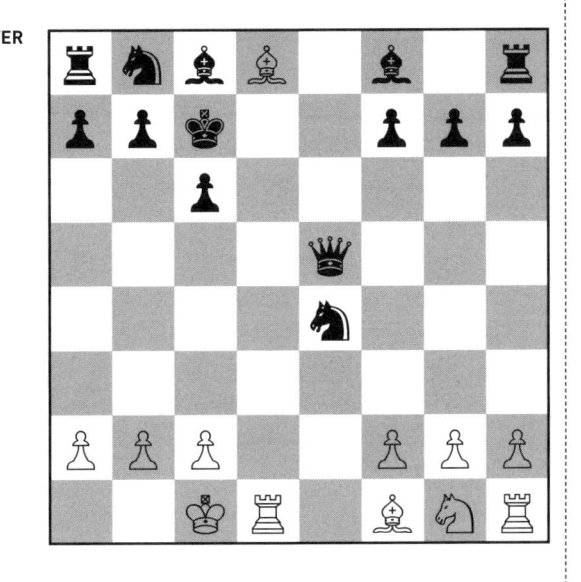

Black's ninth move was forced, but his tenth could have been 10. . . . . K – K1. This would not have affected the result of the game, however, for White would simply have replied 11. R – Q8 mate.

36

Puzzle 16: A golf course has a par of 36 for the first nine holes and for the individual holes in order of play the following pars:

3   3   5   4   4   3   5   5   4

If the tenth hole continued this pattern, what would its par be?

The answer is on page 70.

A dice is a cube with each of the cube's six faces marked. Thus there are 6 x 6 = 36 combinations from throwing two dice. If the dice faces are numbered 1 to 6 and the dice is unbiased, then the probability with one dice of throwing a 4, say, is 1/6 and similarly with throwing a 1, 2, 3, 5 or 6. Few people have any problem with understanding that the probability of throwing a 3 or a 4 with one dice would be 1/6 + 1/6 = 1/3, but what about throwing at least one 3 or 4 if two dice are thrown?

One approach to this question is to note that there are 21 different outcomes when two dice are thrown. Six of these are when the two dice show the same number and 15 are when the two dice show different numbers. The 11 outcomes that show a 3 or a 4 are (1,3), (1,4), (2,3), (2,4), (3,3), (3,4), (4,4), (5,3), (5,4), (6,3) and (6,4). Thus it might be tempting to say that the probability of throwing at least one 3 or 4 if two dice are thrown is 11/21, which is 52%.

This answer cannot be right though, because the probability of not throwing at least one 3 or 4 if two dice are thrown is 4/6 x 4/6 = 16/36, so the probability of throwing at least one 3 or 4 if two dice are thrown is 1 − 16/36 = 20/36 = 56%. This answer is similar to the answer above, but it should be exactly the same. What has gone wrong?

What has 'gone wrong' is counting the 15 outcomes of two dice showing different numbers as 15 possible events when actually there are 30. That is, not adjusting for the fact that throwing a 3 and a 4 for example can be done in two ways: a 3 on the first dice and a 4 on the second, or a 4 on the first dice and a 3 on the second. The 11 outcomes listed near the foot of the facing page that show a 3 or a 4 arise from a total of 20 combinations: (1,3), (1,4), (2,3), (2,4), (3,1), (3,2), (3,3), (3,4), (3,5), (3,6), (4,1), (4,2), (4,3), (4,4), (4,5), (4,6), (5,3), (5,4), (6,3) and (6,4). Hence the answer of 20/36 = 56% is the correct answer.

It was Gerolamo Cardano (1501–76), an Italian mathematician, who first wrote about a systematic treatment of probability. Written around 1564, his book was called *Liber de ludo aleae* (Book on Games of Chance). For him it was not just an academic matter – he was at times a professional gambler. Also, knowing the odds does not just help you win, it also helps identify when you might be being cheated, which clearly was a possibility that Cardano was aware of as his book includes a section on cheating methods. Cardano also did pioneering work with negative numbers and complex numbers.

There is a betting game with three dice that pays 2 to 1 if you throw one six, 3 to 1 if you throw two sixes, and 5 to 1 if you throw three sixes. These seem good odds to non-mathematicians who are new to the game. Their (incorrect) reasoning could be that with one dice you will throw a six on average one-sixth of the time, so by throwing three dice you will get a six on average three-sixths or one-half of the time. With odds averaging better than 2 to 1 it therefore seems you should at least get your money back. You don't – even if the dice are fair (so are not loaded to favour results other than a six).

Out of the 6 x 6 x 6 = 216 combinations from throwing three dice, there will be 5 x 5 x 5 = 125 combinations where there is no six and therefore by subtraction just 216 – 125 = 91 combinations where there is one or more sixes (42%). These 91 are made up of 75 combinations with exactly one six, 15 combinations with exactly two sixes, and just one combination with three sixes. Thus for 216 stakes of $1 you can expect to recover ($2 x 75) + ($3 x 15) + ($5 x 1) = $200, leading to an overall average loss of $16 over 216 games. In this game the return per dollar gambled averages 92.6 cents.

A trick with three nontransitive dice (see page 51) is to allow the other player to choose which dice they want and then to pick the winning dice from the remaining two. You can then expect on average to win five out of every nine throws.

Changing the subject to cubes of a different type, both 36 and 216 are the sums of three consecutive cubes as shown below. Also demonstrated below is that 6 is a perfect number.

$$1 + 2 + 3 = 1 \times 2 \times 3 = 6$$

$$1^3 + 2^3 + 3^3 = 36 = 6^2$$

$$3^3 + 4^3 + 5^3 = 216 = 6^3$$

Answer to Puzzle 12:

For most dominoes their location cannot be ascertained initially because it appears there are two or more places where they could go. For example, it initially appears that the horizontal domino 0-0 could be in row five of the left box or in row five of the right box, but without further investigation we don't know which.

Four dominoes whose positions can immediately be identified unambiguously are 0-4 vertical, 0-6 vertical, 4-4 horizontal and 6-6 horizontal.

Having identified the location of the 0-4 vertical domino we now know that 0-0 horizontal is not in the left box and therefore, as it is the only other possibility, it is in the right box row five. Similarly, we can now identify 0-5 horizontal in the top row of the right box and straight away the 1-1 vertical to the left of it. From here on in the identifying of domino positions is much easier and the full answer is shown below.

4	2	1	4	0	2	2	3
1	5	6	4	0	6	1	3
3	3	6	2	6	3	5	5
6	0	1	2	4	2	3	4
4	0	3	1	1	1	0	0
3	5	4	3	2	5	3	1
3	6	4	0	2	5	4	5

1	0	5	6	0	3	6	1
1	3	3	4	2	2	4	1
4	6	1	1	6	0	2	6
5	2	5	1	6	2	5	5
4	5	0	0	6	6	0	0
2	5	1	3	5	6	0	3
4	4	0	2	4	2	5	6

73 _____

Sheldon Cooper, the lead character in television's *The Big Bang Theory*, claims 73 is the 'best number'. Sheldon says, '73 is the 21st prime number, its mirror 37 is the 12th and its mirror 21 is the product of multiplying, hang on to your hats, 7 and 3.'

Prime numbers are integers excluding 1 that are only divisible without the need for fractions by themselves and one. The number two is the only even prime (all the other even numbers being divisible by two of course). The primes are then 3, 5, 7, 11, 13, 17, 19, 23, etc. That 'etc.' is a bluff, however. As of today, and possibly forever, there is no formula for generating primes, but see below.

The harmonic series comprises the reciprocals of the integers, so is: 1/1, 1/2, 1/3, 1/4 .... It can be demonstrated that the sum of the terms of this series is infinite:

> The second term is one half.

> The sum of the third and fourth terms exceeds 1/4 x 2, so is more than one half.

> The sum of the next four terms exceeds 1/8 x 4, so is more than one half.

> The sum of the next eight terms exceeds 1/16 x 8, so is more than one half.

And so on, meaning there is no value that cannot be beaten by picking enough terms in this series.

Now pretend you are Leonhard Euler, and in the expression 1/1 + 1/2 + 1/3 + 1/4 + ... square each term (multiply each term by itself) as Euler did, making 1 + 1/4 + 1/9 + 1/16 + ... If this does not add up to infinity, then what does it add up to? The answer is $\pi \times \pi / 6$.

The expressions 1/1 + 1/2 + 1/3 + 1/4 ... and 1 + 1/4 + 1/9 + 1/16 ... are special cases of something called the Riemann zeta function, named after Bernhard Riemann (1826–1866). The Riemann hypothesis is about the Riemann zeta function and was the eighth problem in the

23 unsolved problems named by Hilbert in 1900. As a measure of the difficulty in finding a proof for the Riemann hypothesis, Hilbert famously said that, 'if he was to awaken after 500 years of sleep his first question would be: has the Riemann hypothesis been proven?'

What has the Riemann hypothesis got to do with prime numbers? It turns out that the Riemann zeta function can also be expressed using an infinite series of factors involving complex numbers and ascending primes.

Hilbert

So, if the Riemann hypothesis could be proved we might get further insight into how prime numbers progress, which in turn could lead to further excitement for Sheldon.

It is easy to prove that the number of primes is infinite. Assume there are a finite number of primes. Now multiply all of these primes together and then add one. The new number will either be a new prime or have one or more factors that were not included in the finite list of primes. Either way, the assumption that there are a finite number of primes is shown to be false. Euclid in around 300 BC had a proof that was similar to this.

Another proof follows from a theorem proved by the Russian mathematician Pafnuty Chebyshev in 1850. Chebyshev showed that for all integers greater than 1 there is at least one prime number between it and its double. Thus, for example, there is at least one prime between 5 and 10, at least one prime between 10 and 20, at least one prime between 20 and 40, and so on, from which it follows the number of primes is infinite.

Goldbach's Conjecture dates back to 1742. The conjecture is that every even integer above two can be expressed as the sum of two primes. Using computers it has been established that Goldbach's Conjecture is true for integers up to 4 quintillion. An even number greater than 4 quintillion can be expressed as the sum of two odd numbers in more than 1 quintillion different ways, so one could reasonably expect that at least one of those pairs would comprise two primes. However, that is not a proof, and with primes getting rarer as numbers get bigger, Goldbach's Conjecture remains a mere conjecture until such time as a proof is found.

73 is an 'evenly odd number', meaning that it is a number that is one more than a number divisible by four. Because 73 is prime as well as being evenly odd, mathematicians know that:

(a) It can be expressed as the sum of two squares (of integers) and,

(b) This can be done in only one way (for 73 this is 9 + 64).

This is an example of the first half of the two-square theorem proposed by French mathematician Pierre de Fermat (1607–1665), that all evenly odd primes can be expressed as the sum of two squares. The other half of Fermat's two-square theorem is equally remarkable: that any prime that is 'oddly odd' (three more than a number divisible by four) can never, yes never, be expressed as the sum of two squares.

Fermat made a considerable contribution to number theory and famously wrote in the margin of a book he was reading that he had a truly marvellous proof of a theorem that the margin was too narrow to contain. This remark was written around 1637 and not found until Fermat had died. The theorem became known as Fermat's Last Theorem and it is astonishing:

$$\text{If } a^n + b^n = c^n \text{ and } a, b, c \text{ and } n$$
$$\text{are all positive integers,}$$
$$\text{then } n=1 \text{ or } n=2.$$

The n as a superscript means 'to the power of' – see page 22. The theorem says that the sum of two cubes (powers of three) can never be another cube, the sum of two powers of four can never be another power of four, and so on. Fermat's Last Theorem was proved by Andrew Wiles in 1994 using methods that would not have been available to Fermat. Notwithstanding his many other mathematical accomplishments, it is generally supposed that Fermat was mistaken when he thought he had a proof, but we will never know.

Page 24 has a proof that the square root of 2 is irrational. Using Fermat's Last Theorem it is very easy to prove that the cube root of 2 is irrational. We begin by assuming that the cube root of 2 is rational, so can be expressed as A/B. This would then mean that $A^3 = 2B^3 = B^3 + B^3$, which contradicts Fermat's Last Theorem. Thus the cube root of 2 is irrational. The same argument extends to the fourth root of 2, the fifth root of 2, and so on.

If Sheldon had been asked to name a second interesting prime, he might have picked 357,686,312,646,216,567,629,137, a number containing 24 digits. It is a left-truncatable prime, meaning that if you remove any number of digits from the left then the remaining number is still prime.

Thus all of the 24 numbers below are prime:

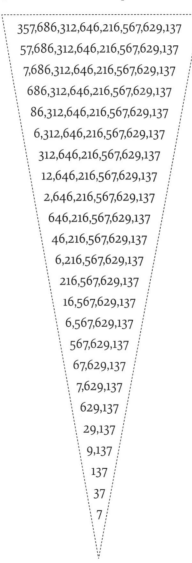

357,686,312,646,216,567,629,137
57,686,312,646,216,567,629,137
7,686,312,646,216,567,629,137
686,312,646,216,567,629,137
86,312,646,216,567,629,137
6,312,646,216,567,629,137
312,646,216,567,629,137
12,646,216,567,629,137
2,646,216,567,629,137
646,216,567,629,137
46,216,567,629,137
6,216,567,629,137
216,567,629,137
16,567,629,137
6,567,629,137
567,629,137
67,629,137
7,629,137
629,137
29,137
9,137
137
37
7

Primes are useful in cryptography (from the Greek for hidden or secret writing) because it is very much easier to multiply two 250-digit primes, say, than to factorize a 500-digit number. Prime numbers can also be used by pseudorandom number generators, which are used in some electronic games.

Puzzle 17: If 73 players enter a single elimination tennis tournament, how many matches excluding byes would it take to decide the winner?

The answer is on page 74.

Answer to Puzzle 11: One answer is that the babies could be two of three triplets.

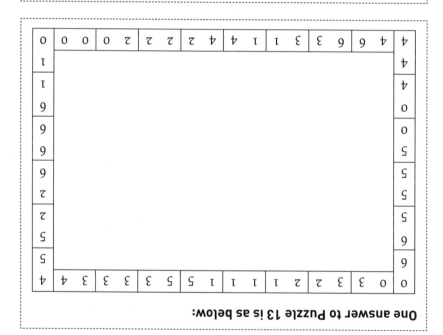

One answer to Puzzle 13 is as below:

Answer to Puzzle 16: The pars of the holes are the number of letters in the hole's number. Thus to maintain the sequence, hole ten would have to be a par 3.

100

There are 100 types of people:

1.	Those who don't understand binary numbers and do get jokes;
10.	Those who don't understand binary numbers and don't get jokes;
11.	Those who do understand binary numbers and don't get jokes;
100.	The ones who get this joke.

Binary numbers only use the digits '0' and '1'; there are no 2s, 3s, 4s, 5s, 6s, 7s, 8s or 9s. Binary numbers are said to be in base 2 and decimal numbers (our everyday numbers) in base ten. Just as 1,834 in decimal means:

$$10^3 \times 1 \quad + \quad 10^2 \times 8 \quad + \quad 10 \times 3 \quad + \quad 1 \times 4$$
$$= 1,000 + 800 + 30 + 4 = 1,834$$

1,101 in binary means:

$$2^3 \times 1 \quad + \quad 2^2 \times 1 \quad + \quad 2 \times 0 \quad + \quad 1 \times 1$$
$$= 8 + 4 + 0 + 1 = 13$$

Thus 10 in binary is 2 in decimal, 11 in binary is 3 in decimal, 100 in binary is 4 in decimal, 1,000 in binary is 8 in decimal and 1,101 is 13 in decimal. Binary numbers are very useful because computers work in binary, essentially by designating 'off' to mean zero and 'on' to mean one.

To distinguish decimal 100 from binary 100 where the context does not make it obvious, mathematicians write these numbers as 100_{10} and 100_2 respectively. Other bases are possible of course and historically were used. Sexagesimal (base 60) was used by the Sumerians close to 6,000 years ago and today we still have 60 minutes in an hour, 60 seconds in a minute, and 360 degrees in a circle. Problem 23 in the Rhind Papyrus (see page 55) is based on the divisors of 360.

Puzzle 18: What is the angle between the hour and minute hands on a clock at twenty past four?

The answer to Puzzle 18 is on page 77.

The ingenuity of the notation '100', whether in binary or decimal, is the use of a symbol that on its own means nothing to mean something. It is the zeros that tell us that the '1' in '100' means something more than 1. We describe our Indo-Arabic number notation as 'positional notation'. The Romans did not have this, which made arithmetic much harder for them. Roman numbers were good for recording, not calculating. Solving the puzzle below will illustrate.

Puzzle 19: In this cross-number, all answers are perfect squares expressed in Roman numerals.

A perfect square is any number that is equal to an integer (whole number) multiplied by itself. Examples of perfect squares are 1, 4, 9, 16, 25 and 36. For the avoidance of doubt in the puzzle above, every across clue and every down clue is the same – 'a perfect square'.

The Roman numerals are: I = 1, V = 5, X = 10, L = 50, C = 100, D = 500 and M = 1,000. Examples of Roman numbers are VIII = 8, LXXVI = 76 and MDCCCLXII = 1862.

The Roman numbers for 4 and 9 are not written as IIII and VIIII, but abbreviated to IV and IX. Similarly with 40, 90, 400 and 900 being abbreviated to XL, XC, CD and CM respectively. Thus 1904 is written as MCMIV and not as MDCCCCIIII.

The logic of the abbreviations is that a numeral to the left of a higher ranking numeral denotes subtraction. It is possible to extend this 'rule' to create other abbreviations such as IL for 49 or XM for 990, but only the six abbreviations mentioned in the paragraph above are available for use in solving this cross-number. Thus if 49 was used in the solution to this puzzle it would be written as XLIX.

The solution to Puzzle 19 is on page 80.

Back to Indo-Arabic numbers, 100 is a 'practical number'. This means that any integer between 1 and 100 can be expressed as the sum of a selection of the factors of 100 using no factor more than once. For example, 83 is the sum of these factors of 100: 50, 25, 5, 2 and 1. Practical numbers are quite common. From 1 to 200 inclusive there are 50 practical numbers. All practical numbers except number 1 are even. Other examples of practical numbers are 12, 42, 128 and 196.

Below is an expression for 100:

$$1 + 3 + 5 + 7 + 9 + 11 + 13 + 15 + 17 + 19 = 100$$

It is the sum of the first 10 odd numbers and not coincidentally 10^2 = 100. It is not coincidental because this works for any sequence of odd numbers beginning with 1. For example, for the first 5 odd numbers, $1 + 3 + 5 + 7 + 9 = 25 = 5^2$. The diagram below illustrates the pattern:

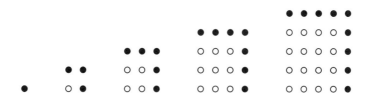

The sum of the first four cubes is also 100, but there is more!

$$1^3 + 2^3 + 3^3 + 4^3 = 100$$

$$(1 + 2 + 3 + 4)^2 = 100$$

This relationship works for any series of integers beginning with 1. Thus

$$1^3 + 2^3 = 9 = (1 + 2)^2$$

$$1^3 + 2^3 + 3^3 = 36 = (1 + 2 + 3)^2$$

$$1^3 + 2^3 + 3^3 + 4^3 + 5^3 = 225 = (1 + 2 + 3 + 4 + 5)^2$$

And so on.

Here is another expression for 100, this time using the nine non-zero digits in ascending order:

$$1 + (2 \times 3) - 4 + (56 \div 7) + 89 = 100$$

It uses six arithmetic signs and brackets for clarity.

> **Puzzle 20: Using no more than the basic arithmetic signs (plus, minus, divide and multiply) and brackets for clarity if needed, find an expression for 100 using the nine non-zero digits in ascending order that uses just three signs and another that uses eight signs.**
>
> **The answers are on page 83.**

Answer to Puzzle 14: The answer is no. Each domino however placed on the chess board will cover a white square and a black square, so 31 dominoes however arranged will cover 31 white squares and 31 black squares. Given that the two white squares removed are of the same colour, thus leaving 30 of one colour and 32 of the other, these 62 squares cannot be covered by 31 dominoes.

Answer to Puzzle 17: To find a winner, 72 players have to be eliminated, so 72 matches are required.

276

Known as D-Day, the World War II cross-channel Allied landings in Normandy on 6 June 1944 were the largest sea invasion in history. Preparation for the invasion had begun the previous year, and part of that preparation was to make an estimate of how many Mark V Panther tanks were being produced by the Germans. Conventional intelligence provided one estimate, but it was the statistical analysis of what has subsequently become known as the 'German Tank Problem' that gave the best result.

To illustrate the process, suppose on a certain tram track the trams in use are numbered sequentially beginning with 1 and three tram numbers have been sighted, the highest being number 30. What is the best estimate of the number of trams using the track?

Given that tram number 30 has been seen, the minimum number of trams is at least 30. The formula for getting the best estimate of the number of trams above the minimum is the minimum number divided by the sample size less one. Thus for the question above, the best estimate of the number of trams using the track is 30 + (30/3 − 1) = 39. Had the highest number sighted been 60 from ten observations, then the estimate would have been 60 + (60/10 − 1) = 65.

For the actual German Tank Problem, the data from captured tanks included serial numbers from engines, gear-boxes, chassis, gun barrels and tyres. The most useful data, however, was derived by estimating the number of wheel moulds in use, because the two captured tanks, each with 32 wheels, supplied 64 data points. The result of the analysis was an estimate of 270 Mark V Panther tanks produced by the Germans in February 1944 against an actual figure (obtained when the war was over) of 276.

A seemingly impossible task is to estimate the size of a population, such as the number of fish in a lake, just by taking samples.

The technique behind the German Tank Problem was applied elsewhere during World War II, with one example being an analysis of the production by the Germans of the V-2 rocket, the world's first long-range ballistic missile. Beginning in September 1944, over a thousand of these were directed at London, killing over 2,000 people. There was at that time no effective defence against a V-2 attack, but through the work of UK statisticians the effectiveness of the new German weapon was reduced.

One of the first jobs of the UK statisticians assigned to the V-2 project was to estimate the accuracy of the V-2. This was done by segmenting London into squares and seeing how many V-2s landed in each square. False information was then 'leaked' saying that the V-2s were over-shooting their targets causing the Germans to (falsely) recalibrate their new weapon. This meant that most V-2s from then on fell short of London into less populated areas. The misinformation to the Germans was maintained by ongoing false reports of large losses of life in London from V-2 missiles.

Solving the German Tank Problem relies on the data collected being numbered in some way. A seemingly impossible task is to estimate the size of a population, such as the number of fish in a lake, just by taking samples. This is how it is done.

Catch a number of fish, mark them, and return them to the lake. When the caught fish have had time to disperse, take another sample and see how many of that sample are marked. Let us suppose it is 10%. If 100 fish were in the first sample, then we can estimate that there were 900 that were not caught for the 10% ratio to apply, giving an answer of

1,000 fish in total in the lake. This statistical technique is known as the capture-recapture method.

Turning to a new topic, amicable numbers come in pairs, the first example being 220 and 284. They are called amicable because the factors of 220 (being 1, 2, 4, 5, 10, 11, 20, 22, 44, 55 and 110 but excluding 220) add up to 284 and the factors of 284 (being 1, 2, 4, 71 and 142 excluding 284) add up to 220. There are over one billion known pairs of amicable numbers. Social numbers are like amicable numbers, except the chain is more than two numbers long. There are fewer than 6,000 known chains of sociable numbers.

A sequence created from an integer by adding up the factors of the previous term, but excluding the term itself, is known as an aliquot sequence. Perfect numbers (see page 60) have a very simple aliquot sequence as by definition the factors of a perfect number add up to the perfect number. Thus the aliquot sequence starting with 28 is 28, 28, 28, 28, ... Numbers where the aliquot sequence ends with an ever repeating perfect number are called aspiring numbers.

Once a prime number appears in an aliquot sequence then the next term is 1 and the terms after that are all zero. Does this mean that all aliquot sequences will eventually repeat with zeros, perfect numbers, or sets of repeating amicable numbers or social numbers? The alternative is that the sequence never repeats. It has not been proved either way, but it seems that the lowest number for starting an aliquot sequence that never repeats is 276.

Answer to Puzzle 18: At twenty past four the minute hand in pointing at 'four' and the hour hand is one third of the way from 'four' to 'five'. The angle between 'four' and 'five' is $360° ÷ 12 = 30°$. One third of $30°$ is $10°$ and that is the answer.

666

This number, according to the Bible (Revelations 13:18), is the number of the beast.

So Puzzle 21 is a story that includes the devil.

A woman is making her way to the Pearly Gates to meet St Peter. In her hand she has a ticket to heaven. In disguise, the devil approaches the woman and tells her he has lost his ticket to heaven and asks the woman if she will share hers. Being a kind woman (she is on her way to heaven), she agrees to share her ticket. With the aid of some divine intervention, the woman takes her ticket and folds A to D and B to C as shown:

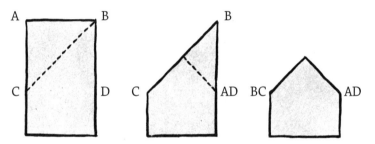

Next, she folded it in half so that BC met AD (let the devil be warned) and marked three sections:

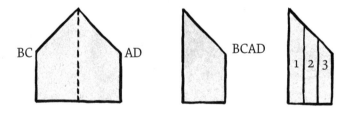

She tears off section three and, after some thought, section two, and gives those pieces to the devil. The devil, ecstatic that his ruse seems to have worked, rushes to the Pearly Gates with his pieces of the ticket to heaven, and hands them over to St Peter demanding admittance. St Peter looks at the pieces, throws them in the air, and this is what they looked like when they landed:

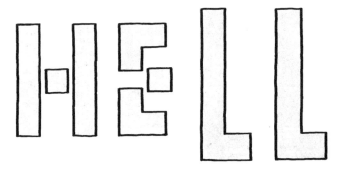

To find out what the residual piece of the ticket to heaven still held by the woman looked like when unfolded, try this yourself. The answer is on page 83.

666 is the value of the first six Roman numerals when written in descending order, being DCLXVI.

Of interest to numerologists, 666 is the sum of the squares of the first seven primes, being $2^2 + 3^2 + 5^2 + 7^2 + 11^2 + 13^2 + 17^2$.

The sum of the numbers on a roulette wheel, which is the sum of the integers from 0 to 36, is 666.

Puzzle 22: Using each of the digits 1, 2, 3, 4 and 5 once and once only and brackets, decimal points and the arithmetic signs of +, - , x and ÷ as required, find an expression that equates to 666. Now do the same for 111, 222, 333, 444, 555, 777, 888 and 999.

The answers are on page 87.

Solution to Puzzle 19:

There are just four two-Roman-digit squares: IV, IX, CD and CM. The Roman numeral I can appear in a Roman number only in the last three positions. Consequently, 2d, 3d, and 8a must all begin with C. By a similar argument for the numerals V and X, 19d will also begin with C.

1a begins with CCC, DCC or MCC. The possibilities are CCCXXIV, CCCLXI, DCCXXIX, DCCLXXXIV, DCCCXLI, MCCXXV, MCCXCVI and MCCCLXIX. Only the last of these is eight numerals long, so 1a is MCCCLXIX.

5d is XXXVI and 6d is XLIX. 14a cannot be IV as then there would be no solution for 15d. 14a is therefore IX and 15d XLIX.

2d and 3d can only be CD or CM. The third and fourth digits of 7a are therefore DM, DD, MM or MD, but cannot be DM or DD. Thus 7a begins with MMMM or MMMD and is therefore MMMDC. Continuing in this vein will solve the puzzle and we can be grateful that today we are using Indo-Arabic numbers!

¹ M	² C	³ C	⁴ C	L	⁵ X	I	⁶ X	
⁷ M	M	M	D	C		X		L
⁸ C	M		⁹ L	X	X	X	I	
¹⁰ C	X	¹¹ L	¹² I	V		V		X
¹³ X	X	X	V	I		¹⁴ I	¹⁵ X	¹⁶ C
¹⁷ X	V	I		¹⁸ M		L		X
V		V	¹⁹ C	D		I		X
	²⁰ M	M	C	M	X	V	I	

711

7-Eleven® is an international chain of convenience stores operating in 17 countries. When the name was created it was in reference to the stores' then opening hours of 7:00 am to 11:00 pm seven days a week, but the name also lends itself to the following puzzle:

> **A customer walks into a 7-Eleven® store and buys four items. The cashier tells the customer that the total cost is $7.11. Curious that the price is the same as the store's name, the customer asks the cashier to check the result as she watches. Put off by being watched, the cashier when using his calculator mistakenly multiplies the four prices instead of adding them. The result is $7.11. They check the result again, this time carefully making sure the prices are added together and not multiplied together. The answer is still $7.11.**
>
> **Telling her friend what happened, the customer was asked what the four prices were. She could not remember, but her friend was able to work them out anyway. What were the four prices?**

This is the solution. To remove the need to use cents, the problem can be expressed as finding four whole numbers that when multiplied together will equal 711,000,000 and when added together will equal 711. The factors of 711,000,000, the integers that when multiplied together will equal 711,000,000, are

2 x 2 x 2 x 2 x 2 x 2 x 3 x 3 x 5 x 5 x 5 x 5 x 5 x 5 x 79.

Having 79 as a factor is helpful because it means that one of the prices must be a multiple of 79 cents. However, with so many other factors of 711,000,000 that can be combined in so many different ways there are still too many possibilities to want to do this puzzle without a computer.

Aided by a computer, we could try out all of the possible price combinations between $0.01 and $7.11 in increments of one cent for each of the four items, but that would entail testing 255,551,481,441 combinations. To reduce the computing time and also to remove duplicate answers, we can instruct the computer without loss of generality that price one will be a multiple of $0.79, price three will be equal to or higher than price two, and price four equal to or higher than price three.

> **Sorting data is an optimization problem that search engines address billions of times a day.**

The answer is the four prices were: $1.20, $1.25, $1.50 and $3.16. There is also a very close, but incorrect, alternative answer. The four prices of $1.01, $1.15, $2.41 and $2.54 sum to $7.11, but these numbers when multiplied together equal 7.1100061, an error of 0.00061 of a cent!

Optimizing computer searches is a science in its own right, and the travelling salesman problem, which was studied as a mathematical problem as far back as 1930, illustrates why. The problem is about a travelling salesman who has a certain number of stop-off points he has to make and who wants to travel the least distance before returning home. With ten stop-offs there are 3.6 million possible routes; with 20 stop-offs there are 2.43 quintillion possible routes, which is a nineteen-digit number. A computer processing a billion routes per second would therefore need 2.43 billion seconds, or 77 years, to test all the options involving just 20 stop-offs.

Sorting data is an optimization problem that search engines address billions of times a day. There are many algorithms (methods) of putting lists in order. Their efficiency relative to one another varies by how sorted the original data is and the number of items to sort. Names of the many techniques available include Bubble Sort, Merge Sort, Heapsort, Insertion Sort and Quicksort. There is even one called Tim Sort, which combines techniques from Merge Sort and Insertion Sort. Data analytics, the analysis of very large amounts of data, is today a career choice for some of the brightest mathematics graduates.

Puzzle 23: A sculpture contains 25 pieces and the weight of each piece is a whole number of kilograms. For any two pieces, the weight of one divided by the weight of the other is never a whole number. What is the lowest possible weight of the sculpture? The answer to Puzzle 23 is on page 87.

Answer to Puzzle 15:

If the question 'What in English is the smallest integer that cannot be defined in less than 28 syllables?' is answered, then that answer has just been defined in less than 28 syllables! This paradox, which can be expressed in many other ways, is known as Berry's Paradox and was first published in 1908 by Bertrand Russell.

Answers to Puzzle 20:

For three signs: 123 − 45 − 67 + 89 = 100.

For eight signs: 1 + 2 + 3 + 4 + 5 + 6 + 7 + (8 × 9) = 100.

Answer to Puzzle 21: Unfolded, the residual piece of the ticket to heaven is shaped as below:

1089

Readers will probably have come across a party trick that typically runs something like this:

> Think of a number.

> Double it.

> Add 10.

> Divide it by two.

> Take away the number you first thought of.

> Drum roll – and the answer is 5.

This is a more sophisticated version:

> Write down a three-digit number using three different digits – call it A.

> Reverse A's digits, call it B, and subtract the smaller of A and B from the other.

> Call the answer C, which is to have three digits, even if the first digit is a zero.

> Reverse the digits of C, call it D, and add D to C.

> Drum roll – and the answer is 1089.

By way of illustration, start with 247:

> The reverse of 247 is 742;
> 742 − 247 = 495;
> The reverse of 495 is 594;
> 495 + 594 = 1089.

And for another example, start with 917:

The reverse is 719;
917 – 719 = 198;
The reverse of 198 is 891;
198 + 891 = 1089.

For the icing on the cake, multiply the answer of 1089 by 9. Reverse the digits of this number (9801) and the answer is, another drum roll, 1089 again.

The reciprocal of 1089 is a repeating decimal with a cycle of 22 digits that for the first 21 digits replicates the nine-times table:

$$\frac{1}{1089} = .00\ 09\ 18\ 27\ 36\ 45\ 54\ 63\ 72\ 81\ 91\ ...$$

This is easily checked as 1089 = 99 x 11 and the reciprocal of 99 is .01 01 01 01 ....

The reciprocal of 9801 is a repeating decimal with a cycle of 198 digits:

$$\frac{1}{9801} = .00\ 01\ 02\ 03\ 04\ 05\ 06\ 07\ 08\ 09\ 10\ 11\ 12\ 13\ 14\ ...\ 94\ 95\ 96\ 97\ 99\ ...$$

For an explanation of why the reciprocal of 9801 is what it is and why there is no 98 see www.timsolepuzzles.com.

For any whole number its 'digital root' is its remainder when it is divided by nine unless it is exactly divisible by nine when its digital root will be nine. Thus the digital root of a number is always an integer from 1 to 9 and a number whose digital root is nine is always divisible by nine.

An easy way of establishing a digital root is to add up the digits in the number. If that answer is greater than nine, then repeat this process with the answer and keep repeating until the digital root is found.

Using 1,234,567 as an example, the sum of its digits is 28, the sum of the digits of 28 is 10, the sum of the digits of 10 is 1, and so the digital root of 1,234,567 is 1. A quicker way to do this is to use a process called 'casting out nines', that is, to remove any nines and any combinations of numbers that add up to nine. In this example, 2 and 7, 3 and 6, and 4 and 5 can be 'cast out' leaving 1 as the digital root.

Digital roots are useful because the digital root of A plus B is the digital root of the digital root of A plus the digital root of B. For example, when 57 is added to 71 the answer is 128, which has a digital root of 2, the same as the digital root of the sum of 3 and 8, being the digital roots of 57 and 71 respectively. Similarly, the digital root of A multiplied by B is the digital root of the digital root of A multiplied by the digital root of B. Digital roots are useful as a quick way to check calculations, but they will not pick up transposition errors.

Looking at the nine times table below:

1 x 9	=	9
2 x 9	=	18
3 x 9	=	27
4 x 9	=	36
5 x 9	=	45
6 x 9	=	54
7 x 9	=	63
8 x 9	=	72
9 x 9	=	81
10 x 9	=	90
11 x 9	=	99
12 x 9	=	108

The digital root of every number on the right is 9, which is as it should be as this is the nine times table. If we now take any one of those numbers and insert one or more zeros or nines, the digital root will still be 9 and the new number will still be divisible by 9. Thus for example, starting with 54, we know that 504, 5004, 5040 and 5400 will all be divisible by 9 without having to do the divisions to find out. Similarly, if we insert threes paired with sixes in any order into any number divisible by nine, making 108 into 16308 or 6160338 for example, the new numbers will also be divisible by 9.

All pan-digital ten-digit numbers (numbers that contain exactly one of each digit) have a digital root of nine, which means that they are all divisible by nine no matter what order the digits appear in. So taking as an example 3,782,915,460, which if the pan-digital numbers were placed in ascending order would be the millionth, it divided by nine is exactly 420,323,940.

Puzzle 24: 1,098,765,432 is a pan-digital number that gives a pan-digital answer when multiplied by 2, 4, 5, 7 or 8. The pan-digital number 8,549,176,320 is pan-digital when divided by 5, but it is more special than that. Can you spot why?

The answer is on page 91.

Answers to Puzzle 22:

$111 = 135 - 24$

$222 = 214 + 3 + 5$

$333 = 345 - 12$

$444 = (152 - 4) \times 3$

$555 = 542 + 13$

$666 = (5 \times 4 \cdot 1 - .2) \div .3$

$777 = (31 \times 5 + .4) \div .2$

$888 = (15^2 - 3) \times 4$

$999 = (5^3 \times 4 \times 2) - 1$

Answer to Puzzle 23: The weight is 711 kilograms. The weights of the individual pieces in kilograms are 8, 12, 14, 17, 18, 19, 20, 21, 22, 23, 25, 26, 27, 29, 30, 31, 33, 35, 37, 39, 41, 43, 45, 47 and 49.

1643

Sir Isaac Newton and Gottfried Leibniz independently invented differential and integral calculus, which makes them very important in the history of mathematics. Sadly there was then a prolonged and bitter dispute with accusations of plagiarism as to who had discovered calculus first. However, these two extraordinarily talented men were each famous for many other things: Leibniz invented the Leibniz wheel, which was used in the first mass-produced calculating machine; Newton in particular for his laws of motion.

Sir Isaac Newton was born on 25 December 1642, so can we say without fear of contradiction that he was born on Christmas Day? Actually no. Newton was born at a time when the Julian calendar was in use. Under the Gregorian calendar that we use today, he was not born on Christmas Day and he was not even born in 1642. He was born on 4 January 1643.

Newton's date of death is even more confusing because new years have not always started on 1 January. When Newton was alive, a new year in England and Wales began on 25 March, but under the Calendar

ISAAC NEWTON

BORN
DECEMBER 25 1642
JANUARY 4 1643

DIED
MARCH 20 1726
MARCH 31 1727

(New Style) Act 1750 (see page 45), 1751 was shortened by 83 days to allow 1752 to begin on 1 January, a change that Scotland had made 152 years earlier. Also under this Act, the year 1700 lost its leap day. Thus although Sir Isaac Newton was recorded as dying on 20 March 1726, we would now call this 31 March 1727.

Serendipity, meaning fortunate happenstance, is a lovely word. It was coined by Horace Walpole in 1754 from the fairy tale *The Three Princes of Serendip*, a tale that features a camel that is lame, blind in one eye and missing a tooth. It was serendipity, an apple falling to the Earth while Newton was in a reflective mood, which caused Newton to think more deeply about gravity. Newton's insight was that if the apple was attracted to the Earth then the Earth might somehow be attracted to the apple and that this force of attraction may still be meaningful even if the objects were a long way apart. In other words, that it could be gravity that was dictating the paths of celestial bodies.

It was also serendipity that a comet appeared in the winter of 1680–81. This would have added extra interest to Newton's research into celestial mechanics (the movements of celestial bodies) that eventually led to Newton's magnum opus, his *Principia Mathematica*, which was published in 1687. *Principia Mathematica* introduced Newton's laws of motion and law of universal gravitation and it was developed from a new type of mathematics that Newton had invented – calculus.

> *Principia Mathematica* introduced Newton's laws of motion and law of universal gravitation and it was developed from a new type of mathematics that Newton had invented – calculus.

'Calculus' is Latin for a 'small pebble used in counting'. In mathematics calculus is the study of change, differential calculus is the study of the rate of change, and integral calculus the accumulation of change. Words like 'the rate of change' or 'the rate of the rate of change' sound complicated, but there are everyday examples of these. For 'rate of change' we might be talking about kilometres per hour; for 'the rate of the rate of change', acceleration.

In a graph of 'y against x', differential calculus provides the gradient (rate of change) of that graph at any point on the graph so long as the formula for calculating 'y' from 'x' meets certain conditions. Integral calculus provides the area underneath the line connecting any two points on such a graph. Thus for a graph that shows speed against time, integral calculus can provide a formula for the distance travelled over a given time period, and differential calculus a formula for the acceleration at any point in time.

To mathematicians' delight, it turns out that integration is the reverse of differentiation and indeed integration is sometimes called anti-differentiation. The fundamental constant of calculus is Euler's number (see page 33), with the differential of 'e to the power of x' (known as the natural exponential function and in mathematical notation written as e^x) being itself.

Calculus was just one of Newton's many mathematical inventions. He also invented a type of sextant (a navigational device) and the Newtonian telescope; he is the Newton of Newton's rings, the Newton scale, the Newton Disc and the well-known Newton's Cradle.

Outside of science (a description Newton would not have agreed with), Newton had a strong interest in studies of the Bible, theology, the occult (which used to mean knowledge of the hidden) and alchemy. On his death, Newton's personal library contained many books on the subject of alchemy and it is thought that in earlier periods of his life he had owned many more. A principal goal of alchemy was to find the 'philosopher's stone', a substance to turn base metals into gold. There were periods in Newton's life where the study of alchemy without official endorsement was punishable by death, so naturally Newton was cautious about what he would publish on this subject.

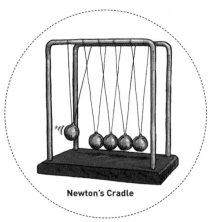

Newton's Cradle

The reason for alchemy being made illegal in Newton's time was not just because promises of a fortune by finding the philosopher's stone were sometimes used as a ruse to swindle people, but because of the concern that if the philosopher's stone was discovered, then that would devalue the gold held by the Crown. Today we scoff at the idea of alchemy, but alchemy was the precursor of chemistry.

In debates as to who over the ages has contributed the most to mathematics and science, it is usually Einstein (see pages 106 and 107) and Newton vying for first position. Readers can decide between these two for themselves, but bear in mind that Newton was developing his theories in calculus and universal gravitation some 200 years before Einstein was born.

The answer to the puzzle that follows is a word that is still in use today, but it would have been more common when Sir Isaac Newton was alive.

Puzzle 25: Which word has kst in the middle, and at the end, and in at the beginning?

The answer to Puzzle 25 is on page 95.

Answer to Puzzle 24: The digits in 8,549,176,320 are in alphabetical order, naming '0' as 'zero'.

1729

Srinivasa Ramanujan (1887–1920), a highly gifted Indian mathematician, lived in England for five years from 1914. G. H. Hardy was a professor at Trinity College Cambridge and also a very gifted mathematician. This anecdote is from when G. H. Hardy visited Ramanujan in hospital. In Hardy's words:

> *I remember once going to see him when he was ill at Putney. I had ridden in taxi cab number 1729 and remarked that the number seemed to me rather a dull one, and that I hoped it was not an unfavourable omen. 'No,' he replied, 'it is a very interesting number; it is the smallest number expressible as the sum of two cubes in two different ways.'*

One of the pairs of cubes is nine and ten cubed and the other pair is one and twelve cubed. Thus:

$$9^3 \quad + \quad 10^3 \quad = \quad 1{,}729 \quad = \quad 1^3 \quad + \quad 12^3$$

Following the theme of taxis, the 'Taxicab Problem':

Puzzle 26 below was devised by Amos Tversky and Daniel Kahneman as part of their research into how people make judgements. This is their question:

A cab was involved in a hit and run accident at night. Two cab companies, the Green and the Blue, operate in the city. 85% of the cabs in the city are Green and 15% are Blue. A witness identified the cab as Blue. The court tested the reliability of the witness under the same circumstances that existed on the night of the accident and concluded that the witness correctly identified each one of the two companies 80% of the time and failed 20% of the time. Knowing that this witness identified the hit and run cab as Blue, what is the probability that the cab involved in the accident was Blue rather than Green?

It is tempting to say that witnesses are right 80% of the time and so the answer is 80%, but that is not the right answer. We can illustrate this by considering our response if we had been told instead that there are 999 Blue cabs for every Green cab, so without the witness's evidence the probability of it being a Blue cab would be 99.9%, and with the witness's evidence it must be more than 99.9%.

The answer to Puzzle 26 is on page 98.

Posterior probabilities, of which the above is an example, can be very confusing to non-statisticians. In legal situations there is even a name for it – the Prosecutor's Fallacy. The theory of conditional probabilities was developed by Thomas Bayes (1701–1761), who first stated the formula for deriving the probability of event B happening if event A has happened from the probability of event A happening if event B has happened.

A well-known problem in conditional probability is the Monty Hall Problem, named after the game show host Monty Hall.

> **This is Puzzle 27:**
>
> You are the winner of a nightly game show and you are being asked to choose your prize. It will be a car or a goat, but you don't know which because the three options being presented to you (two of them are goats) are hidden behind three identical doors. The format is the same every night. Whichever door you choose, the game show host, who already knows what is behind each door, will open one of the other doors to reveal a goat and then ask you if you want to change your mind to the other unopened door. Is it worth changing your mind assuming you want to win the car?
>
> The answer to Puzzle 27 is on page 102.

Perhaps the hardest question in conditional probability is the probability of life itself. As Piet Hien (1905–1996) so eloquently put it:

> *The Universe may*
> *Be as great as they say*
> *But it wouldn't be missed*
> *If it didn't exist*

> **Puzzle 28: Diophantus' epitaph was said to have been (according to a sixth-century book of games and puzzles by Metrodorus):**
>
> *Here lies Diophantus, the wonder behold.*
> *Through art algebraic, the stone tells how old:*
> *God gave him his boyhood, one-sixth of his life,*
> *And one twelfth more until whiskers grew rife;*
> *Another one-seventh ere marriage begun;*
> *In five years came a bouncing new son.*
> *Who died four years before his dad*
> *Who died at the age of twice of his lad's.*
>
> **How old was Diophantus when he died?**
> **The answer to Puzzle 28 is on page 104.**

Diophantus of Alexandria lived in the third century. He was the author of a series of books called 'Arithmetica', one copy of which was the book that Fermat annotated regarding what subsequently became known as Fermat's Last Theorem (see page 68).

Diophantus conjectured that every positive integer can be expressed as the sum of four squares. The four-square theorem was eventually proved in 1770 by Joseph-Louis Lagrange (1736–1813). That led that year to Waring's problem, which asked whether there was a similar theorem for cubes, fourth powers, etc. It was proved that there was by David Hilbert in 1909. A significant contributor to subsequent developments of Waring's problem was Ramanujan's colleague G. H. Hardy.

Waring guessed that the maximum number of cubes of positive integers needed for expressing any integer is nine. It has subsequently been proved that the only numbers needing nine positive cubes are 23 and 229. The largest integer requiring seven positive cubes is 8042 and it is conjectured that all numbers over 1,290,740 can be expressed with five or fewer positive cubes.

87,539,319 is the smallest number that can be expressed by two positive cubes in three different ways:

$$167^3 \ + \ 436^3 \ = \ 228^3 \ + \ 423^3 \ = \ 255^3 \ + \ 414^3 \ = \ 87{,}539{,}319$$

Ramanujan would have liked that.

Answer to Puzzle 25: The word is inkstand, which has 'kst' in the middle, 'and' at the end, and 'in' at the beginning.

5040

Plato (428 BC–348 BC) in his twelve-book work called *Laws* named 5040 as the ideal number of citizens for making up a city. The attractiveness to Plato of 5040 was that it has 60 divisors, including all the numbers from 1 to 10, and when divided into twelfths each portion can then be further divided into twelfths.

5040 is 7 x 6 x 5 x 4 x 3 x 2 x 1, which mathematicians call 'seven factorial' and write as 7!. Henri Brocade in 1885 and Srinivasa Ramanujan independently in 1913 conjectured that there are only three factorials that are one less than a perfect square. It is still not known whether they are right. The known factorials are $4! = 24 = 5^2 - 1$, $5! = 120 = 11^2 - 1$ and $7! = 5040 = 71^2 - 1$.

Multiply together any four consecutive integers and the answer will be one less than the square of the first consecutive integer multiplied by the last consecutive integer plus one. For example:

$$7 \times 8 \times 9 \times 10 \quad = \quad 5040$$

$$(7 \times 10 + 1)^2 \quad = \quad 71^2 \quad = \quad 5041$$

'Five thousand' has no repeated letters and contains each vowel exactly once. 'Forty' is the only number in English whose letters are in alphabetical order. 'One' is the only number in English whose letters are in reverse alphabetical order.

Alphanumerics are puzzles requiring letters to be replaced by digits. These are the usual rules:

> Every letter stands for a different digit;

> If a letter occurs more than once it stands for the same digit;

> Letters at the start of a word cannot represent zero.

Perhaps the best known example of an alphanumeric is SEND + MORE = MONEY, from which S=9, E=5, N=6, D=7, M=1, O=0, R=8, N=6 and Y=2, giving 9567 + 1085 = 10652. Some alphanumerics, including the four in Puzzle 29, require a lot of trial and error and are therefore best solved using a computer search.

Puzzle 29: Four separate alphanumerics, each best solved with a computer search:

S I X	O N E	T W E L V E
S E V E N	T W O	T W E L V E
S E V E N	T W O	T W E L V E
T W E N T Y	T H R E E	T W E L V E
	T H R E E	T W E L V E
	E L E V E N	T H I R T Y
		N I N E T Y

TIM x SOLE = AMOUNT

The answers to Puzzle 29 are on page 102.

Puzzle 30: This alphanumeric can be solved without the need for a computer search:

RYE^3 = INVENTORY

The answer to Puzzle 30 is on page 104.

While we are talking about letters, Zipf's Law, although not a proper law, can be likened to a linguistic equivalent to Benford's Law, which is described on pages 14 to 16. Zipf's Law says that the second most popular word in a language will appear one half as often as the most popular word, the third most popular word one third as often, the fourth most popular one quarter as often, and so on. The most popular word in English is 'the' and the 100 most popular English words account for around 50% of written English.

Other data sets with a pattern similar to Zipf's Law are city populations within countries and the numbers of people within a state or country at any one time watching the same TV channel.

By way of a hint, Puzzle 31 below could be called a cross-wumber puzzle.

1	2	■
■	3	4
5		

Across

1. Starting piece between 3 and 4

3. 2 down without copyright

5. Three!

Down

2. 19 times 4 down

4. Two-thirds or 1.5 times 5 across

The answer to Puzzle 31 is on page 107. If you need a start, 2 down is also the sum of the 17 smallest factors of 5040.

Answer to Puzzle 26: In 100 sightings we would expect 85 x 20% = 17 sightings where a Green cab was identified as Blue and 15 x 80% = 12 sightings where a Blue cab is identified as Blue. Thus in 100 sightings, the witnesses would be expected to say they saw a Blue cab 17 + 12 = 29 times. Of this 29, only 12 out of the 29 were actually Blue, so the probability the witness in the question is correct is 41%.

24752

24752 is my favourite number. It represents 24 hours a day, 7 days a week and 52 weeks a year. It is also my birth date – 24 July 1952. For me, 24752 is not a 'five-digit random number'. Similarly, because of the obvious pattern, I wouldn't choose 12345 or 00000 as five-digit random numbers. These exclusions and similar mean that for me many of the other 99,997 five-digit numbers have a better chance than 1:100,000 of being picked, so they would not be truly random choices either. These other numbers don't stand out because they are interesting; they stand out, albeit not by very much, because they are uninteresting.

This is a true story that illustrates the difficulties of deliberately behaving truly randomly. A man added a porch to his house. He mainly used red bricks, but on instruction from his wife, who was going to be away while he was building the porch, he was to include a small number of white bricks at random. This he did by numbering the red and white bricks using a table of random numbers. On her return the porch has been built and she is not happy. Nearly all of the white bricks had ended up on one side near the top corner. 'I wanted

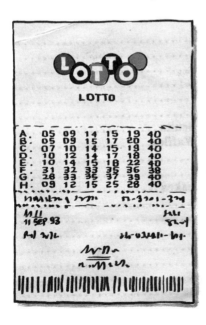

them at random,' she protested, but in fact that is not what she wanted. What she really wanted was for the white bricks to be placed haphazardly.

Series of random numbers are used in statistical trials as benchmarks to test if results are statistically significant, i.e., non-random. They are also used in lotto draws. For lotto, if random numbers produced uninteresting results only, then they wouldn't be random. Here are three examples of interesting lotto results (picked of course from many thousands of seemingly uninteresting results).

To be a Lotto New Zealand jackpot winner a player must pick six winning numbers from 40. The probability of doing this with a single guess is 1 in 3,838,380. In the six months prior to 19 September 2018, Lotto New Zealand averaged 2.3 jackpot winners per draw. On 19 September 2018 it had 40. Prima facie, having 40 jackpot winners in a single draw seems so unlikely that it could never happen. But it did.

The winning numbers (although not the order they were drawn) were 3, 5, 7, 9, 11 and 13, and the 40 jackpot winners won $25,000 each, not the $1,000,000 they had hoped for. How? Many lotto players pick numbers that follow a pattern, and it is not hard to spot the pattern in these winning numbers. It has been suggested that every UK National Lottery draw has approximately 10,000 people who have picked the numbers 1, 2, 3, 4, 5 and 6.

The UK National Lottery on 23 March 2016 did not provide a winner, but it did provide 4,082 players who matched five out of six numbers compared to the usual 50 or so expected. How does that happen? How do 4,082 players get five numbers right, but none get six? The numbers drawn were 7, 14, 21, 35, 41 and 42. Five of these numbers are multiples of seven. It seems likely that many of those 4,082 players who successfully picked five correct numbers had chosen for their six numbers: 7, 14, 21, 28, 35 and 42. These six numbers have the same chance of being chosen at random from the UK lotto's 1 to 59 as any other six, but because many lotto players are not picking numbers at random, the number of winners expected to share the prize pool varies with the six numbers drawn.

The 4,082 players who matched five out of six numbers won £15 each. The 7,879 players who matched just four out of six numbers won £51 each and the 114,232 players who matched three out of six numbers won £25 each. That seems very unfair to the players who matched five out of six, who were actually paid less than those who only matched three or four, but the respective prize pools for three, four and five matches were £61,320, £401,829 and £2,855,800. Had it been just 50 players who guessed five out of six correctly, a more typical result for a draw done by the UK National Lottery, then they would have won £1,226 each instead of £15 each.

Our third example is from the Bulgarian national lottery. Its winning numbers on 6 September 2009 were 4, 15, 23, 24, 35 and 42. No one picked all six, but four days later for the following draw when exactly the same six numbers came up there were 18 winners.

Everyone was very, very surprised of course, but it can be assumed that the majority of the 18 winners would have copied their picks from the previous draw so at least had thought that a repeat was a possibility.

Random numbers are essential for carrying out something called the Monte Carlo method, which has a wide variety of uses. To illustrate, the Monte Carlo method could be used to check for the reasonableness of the birthday statistics on page 56 or the formula for the German Tank Problem on page 75. All that is needed for the Monte Carlo method is a computer, a random number generator, and enough time to run 5,000 simulations (say) to see what the results look like.

π (pronounced pi, see page 37) cannot be expressed as a fraction. As a decimal it has an infinite number of digits. A natural question therefore is whether the digits of the decimal expansion of π can be used as random numbers. Similarly with the digits of the decimal expansions of, amongst others, the golden ratio (see page 25), the square root of 2 (see page 22), the cube root of 2 (see page 68) and Euler's number (e, see page 33).

If the digits of π are random then sequences such as 00000, 12345 and 24752 should each occur about 2000 times in the first 200 million decimal places of π. Using the website www.angio.net/pi they are shown as occurring 1981, 2018 and 1974 times respectively. That is encouraging, but not a proof. As of today, we cannot say for example that the decimal expansion of π contains an unlimited number of zeros. Examining the first 200 million digits of π is impressive, but of course this is only an infinitesimal fraction of the total (infinite) number.

Examining the first 200 million digits of π is impressive, but of course this is only an infinitesimal fraction of the total (infinite) number.

A startling conclusion if π can be treated as an infinite string of random numbers, and mathematicians think this is likely, is that every sequence of numbers will eventually appear and therefore, by applying a code to convert numbers into letters, every written work and everything ever said, and every work that is going to be written and everything that is going to be said … is contained in the digits of π.

Answers to Puzzle 29: The solutions to the three alphanumeric additions are as below:

650	391	130760
68782	803	130760
68782	803	130760
138214	84611	130760
	84611	130760
	171219	194215
		848015

A computer program for solving TWELVE x 5 + THIRTY = NINETY can be found on www.timsolepuzzles.com.

The solution to TIM x SOLE = AMOUNT is 257 x 3806 = 978,142.

Answer to Puzzle 27: If the host had not said anything, then your chance of winning was one third. If having heard what the host said you do not change your mind then your chance of winning is still one third. Therefore, by changing your mind it will be two-thirds. Some people initially think that by following the host's actions the chance of winning a car is 50%, so it does not matter if you swap or not, but that is wrong. Essentially the host has said to you that you can have both of the other two doors and by the way this one has a goat behind it.

142857

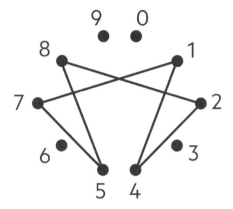

The number 142857 and some of its related numbers below have been written with the digits in different font sizes to emphasize the order of the digits. It and the five new numbers below all follow the sequence in the diagram above.

We begin by doubling 142857 :

$$142857 + 142857 = 285714$$

Taking this answer and doubling it:

$$285714 + 285714 = 571428$$

Subtracting 142857 from this answer:

$$571428 - 142857 = 428571$$

Taking this answer and doubling it:

$$428571 + 428571 = 857142$$

Subtracting 142$857$ from this answer:

$$857{}_{142} - {}_{142}857 = 7{}_{14}285$$

Another pattern:

142,857 x 142,857 = 20,408,122,449: 20,408 + 122,449 = 142,857.

And another pattern:

14 + 28 + 57 = 99: 142 + 857 = 999: 142,857 x 7 = 999,999.

One divided by 7 = 0.142857 142857 142857 142857 …

One divided by 142857 = 0.000007 000007 000007 000007 …

Other numbers with similar properties to 142857 are the first 16 digits of 1 ÷ 17, the first 18 digits of 1 ÷ 19, and the first 22 digits of 1 ÷ 23.

Answer to Puzzle 28: Let a be Diophantus' age when he died, then: a/6 + a/12 + a/7 + 5 + a/2 + 4 = a. Multiply through by 84 to remove the fractions: 14a + 7a + 12a + (5 x 84) + 42a + (4 x 84) = 84a. Which simplifies to: 75a = 84a = 84 x 84 = 75a − 9a. From which a = 84, so Diophantus' age when he died was 84.

Answer to Puzzle 30: E ≠ Y so E ≠ 0, 1, 4, 5, 6 or 9. If E = 2, then Y = 8. The cube of a three-digit number ending in 82 will have as its last two digits '68', so R would be 6. Now we have 682³ = 317,214,568, which works as a solution. Testing the three other possible values of E shows that RYE = 682 is the only solution. When this puzzle was set in an actuarial magazine in February 1980, readers were just told that INVENTORY was a perfect cube (so were not told that RYE was the cube root). Computer run times for those that used a computer search included 11 minutes on a Casio FX-502P and 42 minutes on a Texas Instruments SR 58.

299,792

The family of a friend of mine has two cars. The number plate for one car, which they think is the perfect sports car, is 8128 (a perfect number, see page 60) and the number plate for the other car is EIIMC2. This second number plate is a tribute

to Einstein's famous equation e = mc², where e stands for energy, m for mass, c is the speed of light and c², pronounced 'c squared', means c multiplied by itself. The letter 'c' comes from the abbreviation of the Latin word *celeritas*, meaning swiftness.

Light travels at approximately 299,792 kilometres per second and exactly at 299,792,458 metres per second. This second number is exact because this is how a metre is defined (see page 29).

The speed of light was first estimated in 1676 by Ole Romer, a Danish astronomer. He did it by comparing the apparent fluctuations in the timing of the eclipses of one of Jupiter's moons with the variations in the distance between Jupiter and Earth. Romer calculated the speed of light to be about 220,000 kilometres per second, an error of just 27%. That was quite an achievement given that most astronomers at that time considered light to arrive instantaneously.

In 1883, Simon Newcomb (who also features on pages 13 and 14) in collaboration with the much younger Albert Michelson determined through the use of a spinning mirror that the speed of light was 299,810 kilometres per second; an error margin of less than 0.006%. In 1927, Michelson with better equipment refined that to 299,798 kilometres per second; an error margin of less than 0.002%.

It takes 0.13 seconds for light to travel a distance equivalent to the circumference of the Earth, 8.5 minutes for light to travel from the sun to the Earth, and 5.5 hours for light from the sun to travel to Pluto. It takes 4.4 years for light from the sun to reach Alpha Centauri, the sun's closest star system. The Earth is in the Milky Way, a galaxy that takes light 100,000 years to cross. In the magnitudes that humans think in, light is breathtakingly fast; on a universal scale, with the sizes of galaxies and the distances between them so vast, light is a bit of a plodder!

Mass is a measure of the amount of material in a body and weight is a measure of gravitational force. Thus astronauts in orbit are weightless even though their mass is essentially the same. The mass of the Earth including the atmosphere but excluding the moon is estimated at 5.972 trillion trillion kilograms. A table showing Earth's weight compared to the sun, moon and the other planets in the solar system (Pluto from 2006 being regarded as a dwarf planet) is shown below.

	Mass in Trillion Trillion Kg
Pluto	0.013
Moon	0.073
Mercury	0.33
Mars	0.64
Venus	4.9
Earth	6.0
Uranus	87
Neptune	100
Saturn	570
Jupiter	1,900
Sun	2,000,000

Einstein published his $e = mc^2$ equation in 1905. At that time the concept of energy being a single thing that exists in different forms was barely 50 years old. One form of energy is kinetic energy, a term we use today to describe the energy contained in something by virtue of its speed. Gottfried Leibniz called this energy *vis viva* (living force) and was the first to link it mathematically with energy held by virtue of height, which today we call potential energy. (Example: a cyclist at the top of a hill has potential energy; when speeding at the bottom of a hill he has kinetic energy.)

Another form of potential energy is chemical energy (as per a charged battery). In 1843 James Joule, through a series of experiments, showed that potential energy (through height) had a mechanical equivalent to heat. A year later William Grove postulated that potential energy, kinetic energy, heat, light, magnetism and electricity were a single force (energy in modern parlance). When William Rankine used the phrase 'the law of the conservation of energy' in 1850 he was the first to do so. This was only 55 years prior to the birth of $e = mc^2$.

Measured in metres per second, the speed of light is a nine-digit number and c^2, the speed of light squared, is a 17-digit number. Thus it takes very little mass to generate a huge amount of energy; hence the power of nuclear weapons and the use of nuclear power for generating electricity.

Prior to $e = mc^2$, physicists believed that the conservation of energy and the conservation of mass were distinct, separate laws. Einstein's work in 1905 and his general law of relativity published ten years later overturned these laws as well as superseding Newton's laws of motion. Astonishingly, Einstein's $e = mc^2$ paper in 1905 was just one of the four major works that he published that year. The others were on the photoelectric effect, for which he was awarded a Nobel Prize and which led to quantum mechanics, Brownian motion, and special relativity.

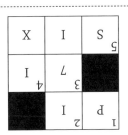

In parts of North America March 14 is known as π Day, being written as 3.14 (see page 37), and is used by some schools as a day for focusing on maths and science projects. Albert Einstein was born on π Day in 1879 and died in 1955.

100,000,000

Archimedes (287 BC–212 BC) was the son of a Greek astronomer, so perhaps it was his father who asked Archimedes how many grains of sand it would take to fill the universe. Never an easy question, but under Greek nomenclature the largest number with a name was a myriad of myriads, which is 100,000,000, and Archimedes knew his answer was going to be bigger than that.

We verbalize 100,000,000 as one hundred million without a second thought. Scientists may opt to say 'ten to the power of eight', which is written as $10^8 = 10 \times 10 \times 10 \times 10 \times 10 \times 10 \times 10 \times 10$. This terminology certainly comes into its own when referring to very large numbers such as a googol, which is 10^{100}. And yes, the company Google did take its name from the number googol and it came from a spelling mistake.

To overcome his counting problem, Archimedes described numbers between 1 and 10^8 as numbers of the first order. He then named a multiple of 10^8, a myriad of myriads, as a number of the second order, and expressed numbers between 10^8 and 10^{16} as a number of the second order plus a number of the first order. He then had numbers of the third order and fourth order and so on. In the same way as today we count in multiples of tens, 523 for example is five hundreds plus two tens plus three, Archimedes for this problem was counting in multiples of a hundred million.

In his paper 'The Sand Reckoner', Archimedes estimated the number of grains of sand that could fit in the universe to be 8 x 10^{63} (8 vigintillion). For this calculation Archimedes assumed in modern terminology that the diameter of the universe was about two light years and that the Earth revolved around the sun. Even putting aside it was done more than 2,000 years ago, it was not a bad estimate. Coincidentally, Eddington's number, the estimated number of nucleons in the observable universe (approximately 47 billion light years in radius), is of the order of 10^{80}, which is roughly equivalent to 10^{63} grains of sand.

Archimedes was both a physicist and mathematician. History records that Archimedes ran naked through the streets shouting 'Eureka' on realizing whilst bathing that he had solved a problem for his king. The problem was to determine, without destroying it, whether the king's new crown was solid gold or gold that was mixed with a metal that was less valuable. For solving the problem with the king's new crown, Archimedes could have compared the volume of the crown with the volume of a block of gold of equal weight by submerging them in water, or he could have weighed the crown in air and then weighed it underwater.

Engraved on Archimedes' tomb at his request was a figure of a sphere and a cylinder to illustrate Archimedes' favourite proof, which was that the volume of a sphere would be two-thirds of that of a cylinder of equal height and diameter. The following puzzle is one that Archimedes would have enjoyed.

Puzzle 32: A cylindrical container of tennis balls is just long and wide enough to hold four tennis balls stored on top of one another. If two tennis balls are removed from the full container the container is then half full. If one tennis ball is then put back, making three tennis balls in the container, can it be said that the container is still half full?

The answer to Puzzle 32 is on page 110.

Archimedes' writings were on plane geometry, geometry in three dimensions, arithmetic, astronomy and mechanics. For the latter, Archimedes' inventions included systems of levers and pulleys and weapons of defence (giant catapults). He is quoted as saying that had he but a fixed fulcrum he could move the Earth. Archimedes also invented the Archimedes Screw, which is still used for irrigation today. It works by turning a handle that operates a spiral tube to lift and move water.

Archimedes is considered one of the greatest mathematicians to have ever lived. Using 96-sided polygons in a process similar to calculus he gave a range for π (see page 37) of 3.1408 to 3.1429, which even at the ends of the range are more than 99.95% accurate. It is an image of Archimedes that features on the Fields Medal, which is widely considered as the Nobel Prize of mathematics.

Fields Medal

Puzzle 33: Without using a computer, find two numbers whose squares add up to 100,000,000.

An answer to Puzzle 33 is on page 115.

Answer to Puzzle 32: Let the volume of the cylinder containing the four tennis balls be four units. Using Archimedes' formula, the volume occupied by one tennis ball is two-thirds of one unit. This means that the volume occupied by three tennis balls is ⅔ x 3 units, which is two units, which is equal to half of the volume of the container. Thus the container when holding three tennis balls can be said to be half full.

PART 3

INFINITY AND BEYOND

Infinity

When I was young my friends and I would play a game where one person would ask another to name the biggest number. Whatever their answer, the response would usually be, 'and that plus one', to which the counter was, 'and turn all the digits of your number into nines'. The first person would then repeat, 'and that plus one', to which

the reply was, 'and turn all the digits of your number into nines'. This could go on for a while, and then someone would say, 'and that plus infinity', to which the reply would be something like, 'plus two more infinities', and so on. It sounds a silly game, but we were curious. What was the biggest number? If it wasn't a line of nines stretching out to who knows where, then what was it?

Is infinity a number? The answers are yes, no, maybe and it depends. (This answer reminds me of a response that was given to the apparently simple question of what is two plus two? Are you buying or selling?) And what of the infinitesimals used in proofs of calculus that George Berkeley in 1734 called 'ghosts of departed quantities'? If infinity is not a number, can an infinitesimal be a number, and if an infinitesimal is not a number, can infinity be?

David Hilbert in a 1924 lecture introduced a concept now referred to as Hilbert's Hotel. It asks how a fully occupied hotel with infinitely many rooms could still accommodate additional guests, even infinitely many of them. Hilbert's descriptions were along the following lines:

To add one more guest: Move the occupier of room one into room two, the occupier of room two into room three, the occupier of room three into room four, and so on. Room one is now free for the new guest.

To add an infinite number of guests: Move the occupier of room one into room two, the occupier of room two into room four, the occupier of room three into room six, and so on, freeing up all the odd-numbered rooms, which is an infinite number.

To add the guests from an infinite number of coaches each holding an infinite number of passengers: Free up the odd-numbered rooms as in the paragraph above, then

> For the first coach, put the first passenger into room 3, the second into room (3 x 3), the third into room (3 x 3 x 3), the fourth into room (3 x 3 x 3 x 3) and so on.

> For coach two, use rooms 5, (5 x 5), (5 x 5 x 5), (5 x 5 x 5 x 5), ...

> For coach three, start with the third odd prime number, which is 7, and use rooms 7, (7 x 7), (7 x 7 x 7), (7 x 7 x 7 x 7), ...

> For coach four, start with the fourth odd prime number, which is 11, and use rooms 11, (11 x 11), (11 x 11 x 11), (11 x 11 x 11 x 11), ...

And so on. Because there are an infinite number of primes (see page 67), everyone in every coach will get their own room.

Stated on the previous page is that there is an infinite number of odd numbers. Even though we know that the integers comprise the odd numbers plus the even numbers, under the mathematics of infinity we say that there is the same number of odd numbers as there are integers.

What do we mean by 'there is the same number?' We mean there is a one-to-one correspondence. The one-to-one correspondence between odd numbers and the integers is illustrated below:

1,   3,   5,   7,   9,   11,   13,   ...

1,   2,   3,   4,   5,   6,   7,   ...

Thus the number of odd numbers is infinite **and** countable. The latter is important, because being countable means we can say that there is a first, second, third and so on, which is what is required to allocate new guests to the rooms in Hilbert's Hotel. For example, as long as the number of new guests is countable, if we wanted to work out exactly which room number the billionth passenger on the millionth coach would be allocated to, we could.

There is a surprising consequence to Hilbert's room shuffling for an infinite number of coaches each holding an infinite number of passengers. Once the rooms have been shuffled round and the new guests accommodated, the hotel that was full now has an infinite number of empty rooms! In addition to room 1, the empty rooms are the rooms whose numbers are odd and which have at least two different prime factors. Examples are 15, 21, 33, 35, 39 and 45.

Puzzle 34: One answer to what is 'never odd or even?' is infinity, but thinking laterally there is a much better answer to this question. What is it?

An answer to Puzzle 34 is on page 119.

Answer to Puzzle 33: $100,000,000 = 100 \times 1,000,000 = (6^2 + 8^2) \times 1,000^2 = 6,000^2 + 8,000^2$.

Beyond Infinity ___

One might think that if infinity plus an infinity of infinities is still infinity, then that would be it, but no. German mathematician Georg Cantor (1845–1918) uncovered a hierarchy of infinites in a short but polarizing paper published in 1874 entitled 'On a Property of the Collection of All Real Algebraic Numbers'.

Cantor's paper distinguished between infinities that were 'countable', such as the set of rational numbers (numbers that can be expressed as one integer divided by another), and those that are not, such as the set of all numbers (rational numbers plus irrational numbers, which together are known as the 'real numbers'). That put Cantor more than 100 years ahead of Buzz Lightyear from the movie *Toy Story*, whose catchphrase is, 'to infinity … and beyond'.

To prove that the rational numbers are countable we just need to demonstrate in the language of Hilbert's Hotel (see previous chapter) that an infinite number of coaches each carrying an infinite number of passengers could hold the rational numbers. In fact, in the seating plan for the coaches that follows, every individual rational number in its own right is allocated an infinite number of seats! The seating plan is for coach one to be allocated the integers divided by one, coach two the integers divided by two, coach three the integers divided by three, and so on. (To illustrate that this seating plan allows every rational number an infinite number of seats, 3/2 for example also appears as 6/4, 9/6, 12/8, etc.)

Before proving the set of real numbers is not countable, a question: which of the two lines below contains the most numbers?

Is it the short line labelled zero to two or the longer one labelled zero to one? Here is the answer:

Firstly, note that a line labelled 0 to 1 can represent a longer distance than a line labelled 0 to 2 if, for example, the units of the 0–1 labelled line are metres and the units of the 0–2 labelled line are feet. However, that doesn't matter here as there is an easy proof that there is the same number of numbers between zero and one as there is between zero and two (or zero and a billion or zero and a tenth). All we need to do is draw two concentric circles, one of which has a circumference of one and the other a circumference of two (or a billion or a tenth).

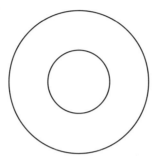

Pick any point on the outer circle and draw a straight line to the circles' centre and that line will cut the inner circle at a unique point. Thus there is the same number of points on the outer circle as there is on the inner circle. We can say this unequivocally because two quantities are defined to be the same if there is a one-to-one correspondence between the items making up those quantities. Thus there are the same number of numbers between zero and one as there are between zero and any other number however big or small that other number is.

To show that the real numbers are not countable, Cantor considered the numbers from zero to one expressed as decimals. For numbers with a finite number of decimals such as 0.25 for example, Cantor substituted 0.249999..., which has an infinite number of decimal places. (For a comparison between 0.9999... and 1 see page 22.)

Cantor's proof begins by assuming the numbers from zero to one are countable, meaning there would be a first, second and third, etc. (It does not matter for the proof what the first, second and third numbers, etc. actually are.) Cantor then constructed a new number that he could demonstrate would not match any of the counted numbers and from this contradiction showed the assumption of countability to be wrong.

Cantor's methodology was as follows:

> Take the digit in the first decimal place of the first number counted and add one. If that gives an answer of ten, then call it zero. Use this as the first decimal place of the new number being constructed.

> Take the digit in the second decimal place of the second number counted and add one. If that gives an answer of ten, then call it zero. Use this as the second decimal place of the new number being constructed.
> Take the digit in the third decimal place of the third number counted and add one. If that gives an answer of ten, then call it zero. Use this as the third decimal place of the new number being constructed.

And so on. This new number is different from the first counted number, is different from the second counted number, and so on. That is, it is different from all the counted numbers, so the assumption that the real numbers are countable is incorrect. This is because however they are counted there will be some that have not been counted. In fact, there will be an uncountable infinity that will not have been counted.

Cantor named the countable infinity as aleph zero, where aleph is the first character of the Hebrew alphabet. It is written as:

Cantor then defined a hierarchy of infinites: aleph zero, aleph one, aleph two, etc. To explain the relationship between these infinities, he used the mathematics of sets and subsets. To illustrate how subsets work, if we take a set of four things such as A, B, C and D there will then be 16 possible subsets, including the empty set, as shown below:

	A	B	C
D	AB	AC	AD
BC	BD	CD	ABC
ABD	ACD	BCD	ABCD

The formula for calculating the number of subsets in this case is 2 x 2 x 2 x 2 = 2^4, because in any subset the first element can be in or out, element two can be in or out, element three can be in or out, and element four can be in or out.

The question then addressed by Cantor was what is the number of subsets from a set containing aleph zero elements such as the set of integers? We can see from the example above that it is two to the power of aleph zero. This is the number that Cantor defined as aleph one. Aleph two is then defined to be the number of subsets of aleph one, and so on.

Such is the strangeness of the maths involving infinity that the number of real numbers despite being an uncountable number can still be quantified! For the possible decimal numbers from zero to one there are ten options for decimal place one, ten for decimal place two, and so on. That means there are 10^∞ of possible decimal numbers from zero to one (and as proved above, from zero to any other number).

Had we done the calculation above in binary (see page 71) using binary decimals our answer would have been 2^∞. Thus 2^∞ and 10^∞ are the same number and more particularly, we have just shown that the number of real numbers in any given range is two to the power of aleph zero, the number that Cantor defined as aleph one.

Cantor proposed his Continuum hypothesis in 1878, which states that there is no number between the number of rational numbers (aleph zero) and the number of real numbers (aleph one). The proof or disproof of this was the first of David Hilbert's 23 problems as described on page 55. Kurt Gödel, an Austrian/American mathematician, in 1938 showed the Continuum hypothesis could not be proved and Paul Cohen, an American mathematician, in 1963 showed it could not be disproved.

Sadly, Cantor and his ideas were subject to very hostile criticism. Beginning in 1884, Cantor had multiple periods in hospital suffering from depression. He was recognized for his work in his lifetime however, receiving the Sylvester Medal from the Royal Society in 1904, its highest honour for a mathematician. Kurt Gödel received the Albert Einstein Award in 1951 and Paul Cohen the Fields Medal in 1966.

Answer to Puzzle 34: 'Never odd or even' is a palindrome – it reads the same forwards and backwards.

GLOSSARY
AND
INDEX

Indo-Arabic Numbers: the notation in use today for writing down numbers. See pages 26 and 72-3.

Integer: a real number that is not a fraction. They are sometimes described as whole numbers.

Irrational Numbers: a number is said to be irrational if it cannot be expressed as the ratio of two integers. See pages 23, 34, 68-9 and 116.

Logarithms/Logs: as an example, log 2 = 0.301 because $10^{0.301}$ (ten to the power of 0.301) = 2. It is convention that log 2 is read as $\log_{10} 2$, the logarithm of 2 in base ten. See pages 13 and 14.

Natural Exponential Function: essentially the graph of y = e^x, where e is Euler's Number. This graph is special because for every value of y the slope of the curve at that point is also y. See Page 34.

Natural Logarithms: logarithms in base e (Euler's Number). See Page 34.

Oddly Odd Numbers: an oddly odd number is a number that is three more than a number divisible by four. See page 68.

Palindrome: something that is the same when read backwards. The famous example (ignoring punctuation and spaces) is, 'A man, a plan, a canal: Panama'. Numbers can also be palindromic, so 101 for example is a palindromic prime.

Square of a Number: the number multiplied by itself, so a number squared is the same as a number to the power of 2. Thus eight squared, which is written as 8^2, is 8 x 8 = 64. See page 22.

Square Root of a Number: a number that when multiplied by itself equals the number. Thus the square root of 100, which is written as $\sqrt{100}$, is either 10 or -10 because 10 x 10 = 100 and -10 x -10 =100. See page 22.

Weird Number: a number whose factors excluding itself total more than the number, but no subset of these factors has a total equal to the number.